METHODS IN MOLECULAR BIOLOGY™

Series Editor
John M. Walker
School of Life Sciences
University of Hertfordshire
Hatfield, Hertfordshire, AL10 9AB, UK

For further volumes:
http://www.springer.com/series/7651

Staphylococcus Epidermidis

Methods and Protocols

Edited by

Paul D. Fey

Department of Pathology and Microbiology, Center for Staphylococcal Research, University of Nebraska Medical Center, Omaha, NE, USA

Humana Press

Editor
Paul D. Fey
Department of Pathology and Microbiology
Center for Staphylococcal Research
University of Nebraska Medical Center
Omaha, NE, USA

ISSN 1064-3745 ISSN 1940-6029 (electronic)
ISBN 978-1-4939-6053-8 ISBN 978-1-62703-736-5 (eBook)
DOI 10.1007/978-1-62703-736-5
Springer New York Heidelberg Dordrecht London

Softcover reprint of the hardcover 1st edition 2014

Printed on acid-free paper

Humana Press is a brand of Springer
Springer is part of Springer Science+Business Media (www.springer.com)

Preface

The chapters in this book are designed to give the new investigator a series of tools to ask novel and exciting questions related to the biology of *Staphylococcus epidermidis* and other staphylococci. We are all indebted to the original pioneers (G. Pulverer, G. Peters, F. Götz, W. Kloos, G. Archer, G. Christensen, K. Schleifer, J. Parisi, and W. Noble among others) who chose to study this commensal and somewhat recalcitrant opportunistic pathogen. These original studies, which primarily focused on the establishment of *S. epidermidis* as an opportunistic pathogen, laid the groundwork for our current understanding of the pathogenesis of *S. epidermidis* including biofilm formation. However, many exciting and unexplored questions remain, such as defining the interaction of *S. epidermidis* and other normal flora. Does *S. epidermidis* have a function in immune system development? Why do certain coagulase negative staphylococci have specific microniches on the human host? What factors function to mediate the binding of *S. epidermidis* to epithelial cells? Due to the highly collaborative nature of investigators working in the field, we have rapidly advanced our understanding of this opportunistic pathogen in the last 20 years. However, due to the commensal relationship between *S. epidermidis* and the human host, I predict that the next 20 years will bring highly novel insights regarding the function of *S. epidermidis* colonization on human development and overall health.

Omaha, NE, USA — ***Paul D. Fey***
May 2013

Contents

Contributors

KENNETH W. BAYLES • *Department of Pathology and Microbiology, Center for Staphylococcal Research, University of Nebraska Medical Center, Omaha, NE, USA*
JEFFREY L. BOSE • *Department of Pathology and Microbiology, Center for Staphylococcal Research, University of Nebraska Medical Center, Omaha, NE, USA*
TESS EIDEM • *Department of Pathology and Microbiology, University of Nebraska Medical Center, Omaha, NE, USA*
JENNIFER L. ENDRES • *Department of Pathology and Microbiology, Center for Staphylococcal Research, University of Nebraska Medical Center, Omaha, NE, USA*
PAUL D. FEY • *Department of Pathology and Microbiology, Center for Staphylococcal Research, University of Nebraska Medical Center, Omaha, NE, USA*
RICHARD V. GOERING • *Department of Medical Microbiology and Immunology, Creighton University School of Medicine, Omaha, NE, USA*
MARK L. HANKE • *Department of Pathology and Microbiology, Center for Staphylococcal Research, University of Nebraska Medical Center, Omaha, NE, USA*
CURTIS W. HARTMAN • *Department of Orthopedic Surgery, University of Nebraska Medical Center, Omaha, NE, USA*
CORTNEY E. HEIM • *Department of Pathology and Microbiology, Center for Staphylococcal Research, University of Nebraska Medical Center, Omaha, NE, USA*
SIOBHAN HOGAN • *Department of Clinical Microbiology, Royal College of Surgeons in Ireland, Dublin, Ireland*
ALEXANDER R. HORSWILL • *Department of Microbiology, University of Iowa, Iowa City, IA, USA*
HWANG-SOO JOO • *Pathogen Molecular Genetics Section, Laboratory of Human Bacterial Pathogenesis, National Institute of Allergy and Infectious Diseases, U.S. National Institutes of Health, Bethesda, MD, USA*
TAMMY KIELIAN • *Department of Pathology and Microbiology, Center for Staphylococcal Research, University of Nebraska Medical Center, Omaha, NE, USA*
KEVIN E. LINDGREN • *Department of Orthopedic Surgery, University of Nebraska Medical Center, Omaha, NE, USA*
JILL K. LINDGREN • *Department of Pathology and Microbiology, Center for Staphylococcal Research, University of Nebraska Medical Center, Omaha, NE, USA*
G. MATTHEW LONGO • *Division of Vascular Surgery, Department of General Surgery, University of Nebraska Medical Center, Omaha, NE, USA*
KATHERINE L. MALISZEWSKI • *Department of Pathology and Microbiology, Center for Staphylococcal Research, University of Nebraska Medical Center, Omaha, NE, USA*
HANNAH MCCARTHY • *Microbiology, School of Natural Sciences, National University of Ireland, Galway, Ireland*
DEREK E. MOORMEIER • *Department of Pathology and Microbiology, Center for Staphylococcal Research, University of Nebraska Medical Center, Omaha, NE, USA*

AUSTIN S. NUXOLL • *Department of Pathology and Microbiology, Center for Staphylococcal Research, University of Nebraska Medical Center, Omaha, NE, USA*
JAMES P. O'GARA • *Microbiology, School of Natural Sciences, National University of Ireland, Galway, Ireland*
EOGHAN O'NEILL • *Department of Clinical Microbiology, Royal College of Surgeons in Ireland, Dublin, Ireland*
MICHAEL E. OLSON • *Department of Microbiology, University of Iowa, Iowa City, IA, USA*
MICHAEL OTTO • *Pathogen Molecular Genetics Section, Laboratory of Human Bacterial Pathogenesis, National Institute of Allergy and Infectious Diseases, U.S. National Institutes of Health, Bethesda, MD, USA*
ROBERT POWERS • *Department of Chemistry, University of Nebraska-Lincoln, Lincoln, NE, USA*
AMITY L. ROBERTS • *Department of Pathology and Microbiology, University of Nebraska Medical Center, Omaha, NE, USA*
D. ASHLEY ROBINSON • *Department of Microbiology, University of Mississippi Medical Center, Jackson, MS, USA*
MARK E. RUPP • *Division of Infectious Disease, Department of Internal Medicine, University of Nebraska Medical Center, Omaha, NE, USA*
CAROLYN R. SCHAEFFER • *Department of Pathology and Microbiology, Center for Staphylococcal Research, University of Nebraska Medical Center, Omaha, NE, USA*
TYLER D. SCHERR • *Department of Pathology and Microbiology, Center for Staphylococcal Research, University of Nebraska Medical Center, Omaha, NE, USA*
JESSICA N. SNOWDEN • *Division of Pediatric Infectious Disease, Department of Pediatrics, University of Nebraska Medical Center, Omaha, NE, USA*
GREG A. SOMERVILLE • *School of Veterinary Medicine and Biomedical Sciences, University of Nebraska-Lincoln, Lincoln, NE, USA*
JONATHAN C. THOMAS • *Department of Microbiology, University of Mississippi Medical Center, Jackson, MS, USA*
ELAINE M. WATERS • *Department of Pathology and Microbiology, Center for Staphylococcal Research, University of Nebraska Medical Center, Omaha, NE, USA*
TODD J. WIDHELM • *Department of Pathology and Microbiology, Center for Staphylococcal Research, University of Nebraska Medical Center, Omaha, NE, USA*
KEITH M. WOODS • *Department of Pathology and Microbiology, Center for Staphylococcal Research, University of Nebraska Medical Center, Omaha, NE, USA*
VIJAY KUMAR YAJJALA • *Department of Pathology and Microbiology, Center for Staphylococcal Research, University of Nebraska Medical Center, Omaha, NE, USA*
MARTA ZAPOTOCZNA • *Department of Clinical Microbiology, Royal College of Surgeons in Ireland, Dublin, Ireland*

Chapter 1

Clinical Characteristics of Infections in Humans Due to *Staphylococcus epidermidis*

Mark E. Rupp

Abstract

Staphylococcus epidermidis is the most common cause of primary bacteremia and infections of indwelling medical devices. The ability to cause disease is linked to its natural niche on human skin and ability to attach and form biofilm on foreign bodies. This review focuses on the *S. epidermidis* clinical syndromes most commonly encountered by clinicians and future potential treatment modalities.

Key words *Staphylococcus epidermidis*, Infection, Bacteremia, Biofilm, Health care-associated infection

1 Introduction

Staphylococcus epidermidis is the most abundant microbe making up the normal commensal flora of human skin and mucous membranes [1]. Due to changes in the practice of modern medicine with increasing use of indwelling prosthetic medical devices and increasing numbers of immunocompromised patients, *S. epidermidis* has become a frequent and formidable pathogen [2, 3]. Presently, *S. epidermidis* is the most common cause of primary bacteremia, is frequently encountered in infections of indwelling medical devices, and is the third most common cause of nosocomial infections [4, 5]. *S. epidermidis* owes its pathogenic success to two major features—its natural niche on human skin, thus resulting in ready access to any device inserted or implanted across the skin, and its ability to adhere to biomaterials and form a biofilm [4, 6–9]. Unfortunately, infections caused by *S. epidermidis* are often indolent and may be clinically difficult to diagnose. Differentiation of culture contamination from true infection can be challenging and treatment is made more difficult by increasing rates of antibiotic resistance and by the effect of biofilms on host defense and antimicrobial susceptibility. Because infected prosthetic devices must

Paul D. Fey (ed.), *Staphylococcus Epidermidis: Methods and Protocols*, Methods in Molecular Biology, vol. 1106,
DOI 10.1007/978-1-62703-736-5_1, © Springer Science+Business Media, LLC 2014

often be removed to exact cure, it is very advisable to put into place rigorous infection prevention measures designed to prevent these infections from occurring in the first place. However, because the chief risk factors for *S. epidermidis* infection—namely the use of indwelling medical devices and the prevalence of immunocompromised hosts, will most likely continue to increase, it is anticipated that the clinical significance of *S. epidermidis* will similarly increase.

2 Colonization and Transmission

Aptly named, *S. epidermidis* is one of the most prevalent species found on human skin, with the average person consistently carrying 10–24 different strains [1, 10]. Depending on the anatomic site, healthy human skin or mucous membranes support from ten to one million colony-forming units (CFU)/cm^2 of *S. epidermidis*. Because of characteristics of human skin, including varying moisture content, nutrient substances, pH range, and temperature, *S. epidermidis* must adapt to a wide range of environmental conditions. For example, to cope with the harsh environmental conditions and high salt concentrations encountered on the dermal surface, *S. epidermidis* possesses a number of osmoprotective systems [3, 4]. Furthermore, most strains of *S. epidermidis* contain the arginine catabolic mobile element (ACME) which is thought to improve the organism's ability to colonize the skin and mucosal membranes [11]. *S. epidermidis* is also well suited to survive in hospital environments. *S. epidermidis* has been noted to survive on fabrics and inanimate surfaces for weeks to months [12] and is often resistant to multiple antibiotics [4]. Decreased susceptibility to certain antiseptics and other biocides is observed in some strains of *S. epidermidis* due to the presence of efflux pumps [13]. *S. epidermidis* produces phenol soluble modulins (PSMs) that are proinflammatory cytolysins and appear to exert selective antimicrobial activity against other organisms, including *S. aureus*, thus helping *S. epidermidis* to outcompete other microbes for dominion of the skin niche [14, 15]. Similarly, *S. epidermidis* serine protease, Esp, has the ability to inhibit biofilm formation and may inhibit nasal colonization by *S. aureus* [16], suggesting that some strains of *S. epidermidis* may in the future be used as probiotics to exclude the more pathogenic *S. aureus* from the skin and nasal mucosa.

Investigators have documented in various patient groups that shortly after admission to the hospital, that their endogenous, usually antibiotic-susceptible, *S. epidermidis* strains are replaced by nosocomial, usually antibiotic-resistant, strains of *S. epidermidis* [17–19]. Predominate clones of *S. epidermidis* often emerge and can persist in hospitals for decades [19, 20]. Health care providers may be carriers of these endemic *S. epidermidis* strains and spread them from patient to patient. These highly successful strains of *S. epidermidis* have been shown to spread from unit to unit, hospital to hospital, and even country to country [21–24].

3 Antibiotic Resistance

S. epidermidis isolated from nosocomial environments are almost always resistant to multiple antimicrobial agents. In two recent large surveillance studies from North America, 73–88 % of isolates were resistant to oxacillin, most strains were resistant to fluoroquinolones and macrolides, and many strains were resistant to clindamycin and trimethoprim–sulfamethoxazole [25, 26]. Similar results were noted in the UK [27]. To date, resistance has rarely been noted to newer agents such as linezolid, daptomycin, quinupristin/dalfopristin, ceftobiprole, televancin, and tigecycline. Although *vanA*-positive coagulase-negative staphylococci have not been isolated to date, strains of *S. epidermidis* with elevated glycopeptide minimal inhibitory concentrations (MICs) have been widely reported [28–31].

Phenotypic expression of methicillin (oxacillin)-resistance in *S. epidermidis* is much more heterotypic than that observed in *S. aureus*, meaning that the percentage of the population that expresses high-level oxacillin resistance is smaller. To address this expression difference, the MIC break point to detect oxacillin resistance is lower for *S. epidermidis* than *S. aureus* (≥0.5 μg/mL vs. >4 μg/mL, respectively) [32]. Regardless of the degree of heterotypy observed, all isolates containing *mecA* (the gene conferring oxacillin resistance) should be regarded as resistant to all β-lactam antibiotics [33]. A particularly onerous feature of *S. epidermidis* is their ability to form biofilms on biomaterials. Tolerance to antibiotics and persister cells is a common theme with *S. epidermidis* growing within a biofilm and these characteristics need to be taken into consideration during treatment [34, 35].

4 Clinical Syndromes

Previously disregarded as nonvirulent contaminants, *S. epidermidis* have more recently been recognized as true pathogens. *S. epidermidis* cause a wide variety of clinical infections, many related to foreign bodies and prosthetic medical devices. A brief discussion of the major types of infections caused by *S. epidermidis* follows.

4.1 Bacteremia and Intravascular Catheter Infections

Coagulase-negative staphylococci are the most common cause and account for approximately 30 % of health care-associated bloodstream infections. Most of these infections are caused by involvement of intravascular catheters or other prosthetic medical devices.

Over 150 million intravascular catheters are used in the USA annually. Although 80,000 vascular catheter infections are estimated to occur in ICUs and hemodialysis units in the USA each year [36], the number of catheter-related bloodstream infections could be as high as 250,000 if patients in other care settings were assessed [37]. Immunosuppressed patients, particularly neonates or those with severe neutropenia, are at increased risk of bloodstream infection. In addition to intravascular catheters, mucosal breakdown caused by cytotoxic chemotherapy may precipitate infection in oncology patients.

The diagnosis of a *S. epidermidis* CVC-related bloodstream infection can be difficult to confirm because most infected CVCs have no obvious local evidence of inflammation or infection [38] and blood isolates of *S. epidermidis* often represent contamination rather than true bacteremia. In the past, a diagnosis of CVC-associated bloodstream infection required the removal of the catheter and semiquantitative or quantitative culture of the catheter tip [39, 40]. Many clinicians now take advantage of automated, continuously monitored blood culture systems and use a 2-h cutoff differential time to positivity assessment on blood cultures drawn from the periphery and the CVC to support the diagnosis of a CVC-associated bloodstream infection [41].

In general, short-term, nontunneled CVCs infected with *S. epidermidis* should be removed as these infections can be associated with sepsis and increased morbidity [42, 43]. In patients with *S. epidermidis* infected tunneled CVCs who do not exhibit signs of severe sepsis, it is permissible to attempt catheter salvage. If the central venous catheter is retained, it is advisable to use systemic antimicrobials through all the lumens of the catheter for 7–14 days [44]. Most health care-associated *S. epidermidis* are methicillin-resistant, and vancomycin is most frequently used to treat these infections. In addition, antimicrobial lock therapy is increasingly employed to treat CVC-related infection with the catheter in situ. Generally, success rates of 80–90 % are noted with a relapse of 20–30 %. If fever or bacteremia persists for more than 3 days after initiating therapy, the CVC should be removed. As mentioned, if the catheter is retained, the clinician should be alert for relapse because this occurs in a substantial minority of patients [45, 46]. Tunnel track or port pocket infections require catheter removal, whereas exit site infections can usually be successfully treated with the CVC in place. Catheter lock therapy can also be used to prevent CVC-associated bloodstream infection [47, 48].

Because *S. epidermidis* occupy such a prominent position in the commensal flora of human skin and mucous membranes, they are frequently encountered as culture contaminants. Approximately 1–6 % of blood cultures are contaminated and coagulase-negative staphylococci—usually *S. epidermidis*, are responsible in 70–80 % of cases [49–51]. Typically, rates of true bacteremia range from 10 to 25 % when *S. epidermidis* are isolated from blood cultures [51, 52]. Determining the clinical significance of *S. epidermidis* isolated from blood cultures is difficult and a variety of clinical and laboratory parameters should be examined when making this determination. This is not a trivial issue, because contaminants treated as true pathogens result in unnecessary antibiotic treatment, emergence of antibiotic resistance, excessive use of laboratory resources, antibiotic-associated side effects and toxicity, and greater expense [53]. Conversely, if true pathogens are disregarded as culture contaminants, the patient suffers from a delay in diagnosis and initiation of effective therapy. In general, blood cultures should be obtained using careful aseptic technique in persons in whom a clinical suspicion for bacteremia exists. Paired cultures should be drawn and, if a central venous catheter (CVC) source is being considered, one of the blood samples should be obtained from the catheter. The site (catheter or periphery) and time of obtainment should be recorded. Patients with clinical signs of sepsis, with multiple positive cultures for *S. epidermidis* that reveal growth in less than 24 h, are much more likely to have true bacteremia. Methods that have been used to prevent blood culture contamination include use of effective skin antiseptics, phlebotomy teams, culture bottle preparation, blood culture kits, and double-needle bottle inoculation [49, 50, 54, 55].

4.2 Endocarditis and Infection of Cardiac Devices and Vascular Grafts

Prosthetic valve endocarditis (PVE), although uncommon, is caused by coagulase-negative staphylococci, usually *S. epidermidis*, in 15–40 % of cases [56–59]. The infection is usually health care-related (resulting from inoculation at the time of surgery) and manifests within 12 months of valve placement. The aortic valve is most frequently involved and patients may present in an acute or more indolent fashion. An acute presentation is characterized by fever and physical evidence of valve dysfunction, whereas peripheral stigmata of endocarditis are more commonly observed in patients exhibiting a more indolent course. The diagnosis is usually confirmed by documenting repeatedly positive blood cultures and vegetations by transesophageal echocardiography. Heart failure occurs in 54 % of cases and more than 80 % have complications, including prosthetic valve dysfunction and intracardiac abscesses. Typically, antibiotic therapy consists of vancomycin and rifampin for at least 6 weeks combined with gentamicin for the first 2 weeks [60]. Isolates susceptible to penicillinase-stable penicillins should be treated with oxacillin or nafcillin instead of vancomycin. There is increasing concern regarding vancomycin "MIC creep" and clinicians are sometimes turning to daptomycin or other vancomycin alternatives. Valve replacement surgery is usually necessary. Despite aggressive therapy, the mortality caused by *S. epidermidis* PVE remains high, at approximately 25 % [58].

Unlike PVE, native valve endocarditis caused by coagulase-negative staphylococci is relatively rare, occurring in only 5–8 % of endocarditis cases [17, 56, 61–63]. Native valve endocarditis is the result of hematogenous seeding of previously damaged heart valves and endocardium. Many of these cases are health care-associated, most often caused by the use of intravascular catheters and cardiac devices, and the causative isolates are usually methicillin-resistant. Prolonged symptoms and physical signs (fever, vascular, or immunologic findings) prior to diagnosis are relatively common [62] and patients often have a very complicated clinical course because of embolic events, rhythm conduction abnormalities, and congestive heart failure [64]. Despite aggressive combined medical–surgical treatment, mortality continues to be approximately 25 %.

Cardiac electrophysiologic device infection (pacemakers, defibrillators) occurs in 1–2 % of device placement procedures and coagulase-negative staphylococci (predominantly *S. epidermidis*) account for approximately 50–60 % of these infections [65–67]. Infection can be limited to the pocket or can spread via intravascular leads to involve endocardial tissue. One fourth of pacemaker infections present acutely within 1 month of insertion, but delays of up to 1–2 years are commonly observed. Diagnosis is generally made by culture of the generator pocket, culture of the device itself, or multiple positive sequential blood cultures with the same strain of *S. epidermidis*. However, only about one-third of patients with implantable cardiac device infection are bacteremic [68, 69]. Transesophageal echocardiography is recommended for all patients and is much more sensitive than transthoracic echocardiography [67, 70]. Successful treatment of implanted electrical cardiac devices generally requires complete removal of the device. Relapse rates and mortality are substantially increased if complete removal of the device is not accomplished [66–68, 70]. Antibiotic therapy typically consists of vancomycin or daptomycin, with or without rifampin, continued for 14 days after device removal for patients with infection limited to the pocket and for 6 weeks for patients with bacteremia, lead involvement, or endocarditis. Device reimplantation, if necessary, should be at a new site when the patient is no longer bacteremic.

Although infection is a relatively rare complication of arterial reconstruction (less than 1–6 %, depending primarily on location of the graft), *S. epidermidis* is one of the most common causes (20–30 %) [71, 72]. The organisms causing these infections are thought to be inoculated at the time of surgery from the patient's skin. Most cases present in an indolent fashion, months to years after surgery and manifest as a false aneurysm, fistula or sinus tract formation, or hemorrhage at the anastomotic site. Diagnosis is usually entertained based on local physical findings and supported by radiographic modalities, such as computed tomography (CT), magnetic resonance imaging (MRI), or ultrasound. Blood cultures are often negative because infection may not extend to the graft lumen. Radiographic guided aspiration of perigraft fluid can be helpful in establishing the diagnosis [73]. Optimum treatment requires a combined medical and surgical approach. Intensive and prolonged antibiotic therapy is important, but surgery is required for cure. "Conservative" therapy with antibiotics and without surgery is associated with a high mortality rate and should be avoided if possible [74].

4.3 Orthopedic Prosthetic Device Infections

S. epidermidis, responsible for 30–43 % of cases, are the most common cause of infection of prosthetic orthopedic devices [75–77]. Causative organisms are presumably inoculated at the time of the arthroplasty and, because of their relatively avirulent nature, may be quite indolent in their clinical presentation. Risk factors for infection that are consistently observed include previous joint surgery, perioperative wound complications, and rheumatoid arthritis. Infections due to *S. epidermidis* usually present 3 months to 2 years postoperatively in an indolent fashion with manifestations of pain at the affected joint, without fever or other systemic symptoms. The diagnosis is supported by the presence of elevated inflammatory markers (erythrocyte sedimentation rate or C-reactive protein) [78]. Radiographic imaging studies may be helpful but are often limited by poor sensitivity and specificity [78, 79]. Aspiration and synovial fluid analysis or tissue biopsy may also be helpful, but can be limited by the localized nature of inflammation and potential contamination. Diagnostic yield is improved by obtaining five or six peri-prosthetic tissue specimens for bacterial culture [80]. The recovery of organisms can be optimized by sonication of the prosthesis at the time of removal [75, 81]. Small colony variants are noted in approximately one-third of prosthetic joint infections caused by *S. epidermidis* [82]. Surgical management of prosthetic joint infections consists of debridement and resection of the prostheses or replacement of the prostheses in one-stage or two-stage procedures. Depending on the type of surgical procedure, antibiotic therapy, usually using vancomycin, lasts from 4 to 6 weeks to up to 6 months [78]. In two-stage procedures, the use of antibiotic spacers may improve outcome. Rifampin is recommended in combination therapy in cases of rifampin-susceptible *S. epidermidis* infection [76]. Success rates of 90–95 % are noted [83]. Prevention strategies include the use of laminar flow operating suites, antimicrobial prophylaxis, and antibiotic-impregnated bone cement.

4.4 Cerebrospinal Fluid Shunt Infections

Infection occurs in approximately 5 % of patients undergoing cerebrospinal fluid shunt implantation. Coagulase-negative staphylococci, most predominantly *S. epidermidis*, are the most common cause and are responsible for approximately one-third to one-half of cases [84]. Risk factors predisposing for infection include age younger than 6 months, shunt revision surgery, scalp dermatitis, duration of the procedure, proficiency of the surgeon, and intraoperative use of a neuroendoscope. Signs and symptoms of shunt infection typically develop within 2 months of shunt insertion and should be suspected in patients with local signs of inflammation, nausea or vomiting, signs of increased intracranial pressure, or shunt malfunction. The diagnosis is confirmed by isolation of *S. epidermidis* from cerebrospinal fluid obtained from the shunt. A modest pleocytosis is usually evident, accompanied by an elevated protein level. Positive blood cultures are observed in patients with infected ventriculoatrial shunts.

Most infections are caused by methicillin-resistant strains of *S. epidermidis*, and combination therapy with vancomycin, gentamicin, and rifampin is a traditional regimen. Vancomycin and gentamicin are often delivered intraventricularly. Rifampin achieves excellent cerebrospinal fluid concentration with systemic administration. Experience with newer anti-staphylococcal agents, such as linezolid or daptomycin, is limited. Successful treatment usually requires shunt removal [85]. In a systematic review and meta-analysis of 17 trials involving 2,134 patients, the administration of perioperative (24 h) prophylactic antibiotics in cerebrospinal fluid shunt surgery significantly decreased the risk of infection [86]. Additional reduction in infection has been achieved through application of strict operative aseptic technique, but use of antibiotic-impregnated shunt catheters has been associated with inconsistent results [87].

4.5 Surgical Site Infections

Surgical site infections caused by *S. epidermidis* occur frequently, being second only to *S. aureus* as a causative agent [5, 88]. *S. epidermidis* are more often causative of superficial incisional infections rather than deep incisional infections and rarely cause organ or space infections. A notable exception is mediastinitis after median sternotomy for cardiac surgery. Superficial incisional infections generally manifest within 5–10 days postprocedure and usually result from inoculation of organisms from the patient's endogenous flora or, less frequently, from the operating personnel or environment. Risk factors for development of surgical site infection due to *S. epidermidis* include duration of the surgical procedure, host factors, and experience of the surgeon and surgical staff. Signs and symptoms of a surgical site infection include pain, tenderness, swelling, warmth, erythema, drainage at the incisional site, leukocytosis, and fever. The causative pathogen is confirmed by recovery of *S. epidermidis* from wound cultures. Culture results require careful interpretation because coagulase-negative staphylococci are commonly regarded as contaminants or colonizers. Generally, *S. epidermidis* are interpreted to be the causative agent of the infection if they are the predominant or only isolate from purulent drainage and/or are repeatedly cultured from the same source.

Treatment depends on the severity of the infection and ranges from topical wound care alone to surgical debridement and parenteral antibiotics.

4.6 Peritoneal Dialysis Catheter-Associated Infections

S. epidermidis is responsible for 20–40 % of cases of peritonitis in patients undergoing peritoneal dialysis [89]. *S. epidermidis* gain access to the peritoneum, usually from the patient's skin, via the intraluminal route or from the exit site via the periluminal route. Clinically, compared with infection by *S. aureus*, *S. epidermidis* peritonitis is relatively benign and infrequently leads to catheter removal [90]. Signs and symptoms of infection include abdominal pain and tenderness, fever, nausea, and vomiting. Diagnosis is confirmed by documenting more than 100 white blood cells/mL in dialysate fluid and recovery of *S. epidermidis* in culture of the fluid. Treatment with vancomycin via the dialysate fluid is a relatively convenient dosing method and is often successful. Recalcitrant or recurrent peritonitis is an indication for catheter removal [91].

4.7 Endophthalmitis

Coagulase-negative staphylococci are readily recovered from conjunctival cultures of preophthalmologic surgery patients [92], and thus it is not surprising that *S. epidermidis* is the most frequent cause of postoperative endophthalmitis [93]. Symptoms typically develop within 1 week of surgery and usually consist of pain, redness, and decreased visual acuity. Fever is generally absent and leukocyte count is often normal. The physical examination usually reveals conjunctival injection and a hypopyon. Optimal treatment consists of vitrectomy and intravitreal administration of antibiotics [94]. A large proportion of *S. epidermidis* causing eye infections are multi-drug resistant and vancomycin is usually administered at an intravitreal dose of 1 mg and bactericidal concentrations usually persist for 2–3 days [95]. Intraocular lens removal is generally not required. Although prognosis largely depends on presenting visual acuity, residual visual impairment is frequently observed.

4.8 Infections of Genitourinary Prostheses

S. epidermidis is responsible for 35–60 % of infections of synthetic urinary sphincters and penile prostheses in which an overall infection rate of 2–4 % is observed [96]. Infections of penile prostheses are often indolent and may take up to a year from the date of implantation to manifest clinically. Those with infected prostheses exhibit local pain, swelling, induration, and erythema of the penis. Surgical removal of the device is required, accompanied by 10–14 days of systemic antibiotics for uncomplicated infection [97].

4.9 Mastitis and Infections of Breast Implants

Although *S. aureus* is the major etiologic agent of mastitis in lactating women, *S. epidermidis* may also play a role [98]. Conversely, *S. epidermidis* is a relatively common cause of infection of breast implants. Infection may present acutely or may be very indolent. Signs and symptoms are predominantly localized and include erythema, tenderness, pain, swelling, induration, and drainage. Acute infections are often associated with systemic findings such as fever and leukocytosis. Diagnosis is confirmed by culture of the drainage or fluid surrounding the implant or of the implant itself. Treatment consists of antibiotic therapy and a two-stage replacement procedure. Capsular contracture remains the most common complication following breast augmentation. Although controversial, chronic low-grade or subclinical infection with *S. epidermidis* may be a cause for capsular contracture [99].

4.10 Miscellaneous Prosthetic Device Infections, Implant Infections, and Other Infections

Almost any biomaterial or device that is inserted or implanted across the skin or mucous membranes can become colonized or infected by *S. epidermidis*. Miscellaneous devices that have been associated with infections caused by *S. epidermidis* include ventricular assist devices, coronary stents, hemodialysis shunts and catheters, implantable neurologic stimulators and peripheral nerve catheters, cochlear implants, fracture fixation devices and other orthopedic implants, ureteral or urethral stents, and surgical mesh. It can be anticipated that infection caused by *S. epidermidis* will parallel the increasing use of such devices [97, 100]. In recent years, a large number of infections, including otitis media, have been characterized as biofilm-associated. Thus, it is not surprising that *S. epidermidis*, whose chief virulence determinant is its ability to elaborate a biofilm, have been incriminated as a cause of otitis [101]. *S. epidermidis* have also recently been associated with Rosacea [102].

5 Patient Populations at Increased Risk of Infection by *S. epidermidis*

Solid organ or hematopoietic stem cell transplant patients are susceptible to *S. epidermidis* infections due to immunosuppression, intravascular catheterization, and mucosal or skin breakdown [103, 104]. These infections most often manifest as a bloodstream infection and are due to an infected intravenous catheter. Mucositis and the breakdown of gastrointestinal mucosal integrity, related to cytotoxic chemotherapy or radiation therapy, may be an alternate source for *S. epidermidis* bacteremia [105, 106].

Approximately 20 % of very low birth weight preterm infants (less than 1,500 g) experience late-onset neonatal sepsis (more than 3 days after birth) [107, 108]. Half of these infections are caused by coagulase-negative staphylococci most commonly (60–93 %) *S. epidermidis*. Neonates become colonized with *S. epidermidis* on their skin and in their nares, umbilicus, pharynx, and gastrointestinal tract within days of their admission to the neonatal intensive care unit (NICU) and, in most cases, these organisms do not originate from the mother but instead are acquired from the hospital environment and health care workers [109]. Endemic strains of *S. epidermidis* can persist in the NICU for many years [19].

S. epidermidis bacteremia is often indolent and signs of infection may include abdominal distention, apnea, bradycardia, inability to maintain body temperature, feeding difficulties, lethargy, neutropenia, thrombocytopenia, hyperglycemia, and metabolic acidosis [107]. Differentiating true bacteremia from contamination is made more difficult in neonates for several reasons, including the difficulty in obtaining blood from low-birth-weight infants, the small volume of blood generally obtained (0.1–1 mL), and the common practice of obtaining a single sample of blood for culture to preserve blood volume. *S. epidermidis* infections in the neonate infrequently result in mortality but are often associated with morbidity requiring additional days of care in the hospital while receiving antimicrobial therapy. Prevention of *S. epidermidis* infection in neonates has largely concentrated on prevention of intravascular catheter-associated infection.

Not surprisingly, burn patients have an exceptionally high rate of bloodstream infection. The rate of central-line associated bloodstream infection is higher in burn units (3.7 per 1,000 CVC days) than any other surveyed unit and *S. epidermidis* are among the most frequently isolated pathogens [110, 111]. These infections are associated with increased morbidity and length of hospital stay which emphasize the need for scrupulous attention to CVC insertion and care guidelines.

6 Novel Therapeutic Options for Infections Due to *S. epidermidis*

A greater understanding of the pathogenesis of *S. epidermidis* infections has prompted interest in the development of novel therapeutic modalities [112]. Investigational agents, many of which are thought to interfere with *S. epidermidis* biofilm, include cinnamon oil [113], cathelicidin [114], berberine [115], furanones [116], and pleuromutilins [117]. Biomedical engineers continue to work on biomaterials and prosthetic devices that inhibit staphylococcal adherence or proliferation [118]. Antibiofilm approaches and other innovative technologies include immunoprophylaxis (*S. epidermidis* vaccine) [119], quorum sensing interference, impairment of *S. epidermidis* adhesion or biofilm accumulation, immunotherapy (bacteriophage), enzymatic disruption or removal of biofilm [120], immunomodulation, and use of nanoparticles to deliver anti-biofilm agents. Finally, investigators are attempting to better define the anti-biofilm activity that has been associated with some microbes or parasites [121, 122]. Although some of these developments appear promising, at present, *S. epidermidis* continues to offer vexing challenges to patients, clinicians, and medical microbiologists. Unfortunately, with the increased use of medical devices, the challenge of infection due to *S. epidermidis* is not likely to diminish.

References

1. Grice EA, Kong HH, Conlan S et al (2009) Topographical and temporal diversity of the human skin microbiome. Science 324: 1190–1192
2. Rupp ME, Archer GL (1994) Coagulase-negative staphylococci: pathogens associated with medical progress. Clin Infect Dis 19: 231–243
3. Otto M (2009) *Staphylococcus epidermidis*—the 'accidental' pathogen. Nat Rev Microbiol 7:555–567
4. Rogers KL, Fey PD, Rupp ME (2009) Epidemiology of infections due to coagulase-negative staphylococci. In: Crossley KB, Jefferson KK, Archer G, Fowler VG Jr (eds) The Staphylococci in human disease, 2nd edn. Blackwell Publishing, Oxford, pp 310–332
5. Sievert DM, Ricks P, Edwards JR et al (2013) Antimicrobial-resistant pathogens associated with healthcare-associated infections: summary of data reported to the National Healthcare Safety Network at the Centers for Disease Control and Prevention, 2009–2010. Infect Control Hosp Epidemiol 34:1–14
6. Mack D, Horstkotte MA, Rhode H et al (2006) Coagulase-negative staphylococci. In: Pace JL, Rupp ME, Finch RG (eds) Biofilms, infection, and antimicrobial therapy. Taylor & Francis, Boca Raton, FL, pp 109–153
7. O'Gara JP, Humphreys H (2001) *Staphylococcus epidermidis* biofilms: importance and implications. J Med Microbiol 50: 582–587
8. von Eiff C, Peters G, Heilmann C (2002) Pathogenesis of infections due to coagulase-negative staphylococci. Lancet Infect Dis 2: 677–685
9. Bannerman TL, Peacock SJ (2007) Staphylococcus, Micrococcus, and other catalase-positive cocci. In: Murray PR, Baron EJ, Jorgensen JH et al (eds) Manual of clinical microbiology. American Society for Microbiology, Washington, DC, pp 390–411
10. Kloos WE, Musselwhite MS (1975) Distribution and persistence of *Staphylococcus* and *Micrococcus* species and other aerobic bacteria on human skin. Appl Microbiol 30:381–385
11. Barbier F, Lebeaux D, Hernandez D et al (2011) High prevalence of the arginine catabolic mobile element in carriage isolates of methicillin-resistant *Staphylococcus epidermidis*. J Antimicrob Chemother 66:29–36
12. Neely AN, Maley MP (2000) Survival of enterococci and staphylococci on hospital fabrics and plastics. J Clin Micro 38:724–726
13. Sidhu MS, Heir E, Leegaard T et al (2002) Frequency of disinfectant resistance genes and genetic linkage with B-lactamase transposon Tn*552* among clinical staphylococci. Antimicrob Agents Chemother 46: 2797–2803
14. Frank DN, Feazel LM, Bessesen MT et al (2010) The human microbiota and *Staphylococcus aureus* carriage. PLoS One 5:e10598
15. Cogen AL, Yamasaki K, Sanchez KM et al (2010) Selective antimicrobial action is provided by phenol-soluble modulins derived from *Staphylococcus epidermidis*, a normal resident of the skin. J Invest Dermatol 130:192–200
16. Iwase T, Uehara Y, Shinji H et al (2010) *Staphylococcus epidermidis* Esp inhibits *Staphylococcus aureus* biofilm formation and nasal colonization. Nature 20:346–349
17. Ahlstrand E, Persson L, Tidefelt U (2012) Alteration of the colonization pattern of coagulase-negative staphylococci in patients undergoing treatment for hematologic malignancy. Eur J Clin Microbiol Infect Dis 31: 1679–87
18. Zingg W, Demartines N, Imhof A et al (2009) Rapid colonization with methicillin-resistant coagulase-negative staphylococci after surgery. World J Surg 33:2058–2062
19. Huebner J, Pier GB, Maslow JN et al (1994) Endemic nosocomial transmission of *Staphylococcus epidermidis* bacteremia isolates in a neonatal intensive care unit over 10 years. J Infect Dis 169:526–531
20. Krediet TG, Mascini EM, vanRooij E et al (2004) Molecular epidemiology of coagulase-negative staphylococci causing sepsis in a neonatal intensive care unit over an 11-year period. J Clin Microbiol 42:992–995
21. Kozitskaya S, Olson ME, Fey PD et al (2005) Clonal analysis of *Staphylococcus epidermidis* isolates carrying or lacking biofilm-mediating genes by multilocus sequence typing. J Clin Microbiol 43:4751–4757
22. Widerstrom M, Monsen T, Karlsson C et al (2006) Molecular epidemiology of meticillin-resistant coagulase-negative staphylococci in a Swedish county hospital: evidence of intra- and interhospital clonal spread. J Hosp Infect 64:177–183
23. Gordon RJ, Miragaia M, Weinberg AD et al (2012) *Staphylococcus epidermidis* colonization is highly clonal across US cardiac centers. J Infect Dis 205:1391–1398
24. Widerstrom M, Monsen T, Karlsson C et al (2009) Clonality among multidrug-resistant hospital-associated *Staphylococcus epidermidis* in northern Europe. Scand J Infect Dis 41: 642–649
25. Streit JM, Jones RN, Sader HS et al (2004) Assessment of pathogen occurrences and

resistance profiles among infected patients in the intensive care unit: report from the SENTRY Antimicrobial Surveillance Program (North America, 2001). Int J Antimicrob Agents 24:111–118
26. Jones RN, Ross JE, Castanheira M et al (2008) United States resistance surveillance results for linezolid (LEADER Program for 2007). Diagn Microbiol Infect Dis 62:416–426
27. Hope R, Livermore DM, Brick G et al (2008) Non-susceptibility trends among staphylococci from bacteraemias in the UK and Ireland, 2001–06. J Antimicrob Chemother 62(Suppl 2):ii65–ii74
28. Sieradzki K, Villari P, Tomasz A (1998) Decreased susceptibilities to teicoplanin and vancomycin among coagulase-negative methicillin-resistant clinical isolates of staphylococci. Antimicrob Agents Chemother 42:100–107
29. Sieradzki K, Roberts RB, Serur D et al (1999) Heterogeneously vancomycin-resistant *Staphylococcus epidermidis* strain causing recurrent peritonitis in a dialysis patient during vancomycin therapy. J Clin Microbiol 37:39–44
30. Nunes AP, Teixeira LM, Iorio NL et al (2006) Heterogeneous resistance to vancomycin in *Staphylococcus epidermidis, Staphylococcus haemolyticus* and *Staphylococcus warneri* clinical strains: characterisation of glycopeptide susceptibility profiles and cell wall thickening. Int J Antimicrob Agents 27:307–315
31. Froggatt JW, Johnston JL, Galetto DW et al (1989) Antimicrobial resistance in nosocomial isolates of *Staphylococcus haemolyticus.* Antimicrob Agents Chemother 33:460–466
32. Clinical and Laboratory Standards Institute (2013) Performance standards for antimicrobial susceptibility testing: twenty-third information supplement. CLSI Document M100–S23. Clinical and Laboratory Standards Institute, Wayne, PA
33. Vazquez GJ, Archer GL (1980) Antibiotic therapy of experimental *Staphylococcus epidermidis* endocarditis. Antimicrob Agents Chemother 17:280–285
34. Lewis K (2007) Persister cells, dormancy and infectious disease. Nat Rev Microbiol 5:48–56
35. Fey PD (2010) Modality of bacterial growth presents unique targets: how do we treat biofilm-mediated infections? Curr Opin Microbiol 13:610–615
36. Srinivasan A, Wise M, Bell M et al (2011) Vital signs: central line-associated blood stream infections—United States 2001, 2008, and 2009. MMWR Morb Mortal Wkly Rep 60:243–248
37. Maki DG, Kluger DM, Crnich CJ (2006) The risk of bloodstream infection in adults with different intravascular devices: a systematic review of 200 published prospective studies. Mayo Clin Proc 81:1159–1171
38. Hewlett AL, Rupp ME (2012) Healthcare-associated infections related to the use of intravascular devices inserted for short-term vascular access. In: Mayhall CG (ed) Hospital epidemiology and infection control, 4th edn. Wolters Kluwer, Philadelphia, pp 241–247
39. Sherertz RJ, Heard SO, Raad II (1997) Diagnosis of triple-lumen catheter infection: comparison of roll plate, sonication, and flushing methodologies. J Clin Microbiol 35:641–646
40. Maki DG, Weise CE, Sarafin HW (1977) A semiquantitative culture method for identifying intravenous catheter–related infection. N Engl J Med 296:1305–1309
41. Raad I, Hanna HA, Alakech B et al (2004) Differential time to positivity: a useful method for diagnosing catheter-related bloodstream infections. Ann Intern Med 140:18–25
42. Lebeaux D, Larroque B, Gellen-Dautremer J et al (2012) Clinical outcome after a totally implantable venous access port-related infection in cancer patients. Medicine 91:309–318
43. Molina J, Penuela I, Lepe J et al (2013) Mortality and hospital stay related to coagulase-negative Staphylococci bacteremia in non-critical patients. J Infect 66:155–162
44. Mermel LA, Allon M, Bouza E et al (2009) Clinical practice guidelines for the diagnosis and management of intravascular catheter-related infection: 2009 update by the Infectious Diseases Society of America. Clin Infect Dis 49:1–45
45. O'Horo JC, Silva GLM, Safdar N (2011) Anti-infective locks for treatment of central line-associated bloodstream infection: a systematic review and meta-analysis. Am J Nephrol 34:415–422
46. Raad I, Davis S, Khan A et al (1992) Impact of central venous catheter removal on the recurrence of catheter-related coagulase-negative staphylococcal bacteremia. Infect Control Hosp Epidemiol 13:215–221
47. O'Grady NP, Alexander M, Burns LA et al (2011) Summary of recommendations: guidelines for the prevention of intravascular catheter-related infections. Clin Infect Dis 52:1087–1099
48. Snaterse M, Ruger W, Scholte WJM et al (2010) Antibiotic-based catheter lock solutions for prevention of catheter-related bloodstream infection: a systematic review of randomized controlled trials. J Hosp Infect 75:1–11

49. Weinstein MP, Towns ML, Quartey SM et al (1997) The clinical significance of positive blood cultures in the 1990s: a prospective comprehensive evaluation of the microbiology, epidemiology, and outcome of bacteremia and fungemia in adults. Clin Infect Dis 24:584–602
50. Hall KK, Lyman JA (2006) Updated review of blood culture contamination. Clin Microbiol Rev 19:788–802
51. Pien BC, Sundaram P, Raoof N et al (2010) The clinical and prognostic importance of positive blood cultures in adults. Am J Med 123:819–828.132
52. Souvenir D, Anderson DE Jr, Palpant S et al (1998) Blood cultures positive for coagulase-negative staphylococci: antisepsis, pseudobacteremia, and therapy of patients. J Clin Microbiol 36:1923–1926
53. Van der Heijden YF, Miller G, Wright PW et al (2011) Clinical impact of blood cultures contaminated with coagulase-negative staphylococci at an academic medical center. Infect Control Hosp Epidemiol 32:623–625
54. Spitalnic SJ, Woolard RH, Mermel LA (1995) The significance of changing needles when inoculating blood cultures: a meta-analysis. Clin Infect Dis 21:1103–1106
55. Schifman RB, Strand CL, Meier FA et al (1998) Blood culture contamination: a College of American Pathologists Q-Probes study involving 640 institutions and 497134 specimens from adult patients. Arch Pathol Lab Med 122:216–221
56. Demitrovicova A, Hricak V, Karvay M et al (2007) Endocarditis due to coagulase-negative staphylococci: data from a 22-years national survey. Scand J Infect Dis 39:655–656
57. Lalani T, Kanafani ZA, Chu VH et al (2006) Prosthetic valve endocarditis due to coagulase-negative staphylococci: findings from the International Collaboration on Endocarditis Merged Database. Eur J Clin Microbiol Infect Dis 25:365–368
58. Wang A, Athan E, Pappas PA et al (2007) Contemporary clinical profile and outcome of prosthetic valve endocarditis. JAMA 297: 1354–1361
59. Lee JH, Burner KD, Fealey ME et al (2011) Prosthetic valve endocarditis: clinicopathological correlates in 122 surgical specimens from 116 patients (1985–2004). Cardiovasc Pathol 20:26–35
60. Baddour LM, Wilson WR, Bayer AS et al (2005) Infective endocarditis: diagnosis, antimicrobial therapy, and management of complications: a statement for health care professionals from the Committee on Rheumatic Fever, Endocarditis, and Kawasaki Disease, Council on Cardiovascular Disease in the Young, and the Councils on Clinical Cardiology, Stroke, and Cardiovascular Surgery and Anesthesia, American Heart Association: endorsed by the Infectious Diseases Society of America. Circulation 11:e394–e434
61. Agvald-Ohman C, Lund B, Edlund C (2004) Multiresistant coagulase-negative staphylococci disseminate frequently between intubated patients in a multidisciplinary intensive care unit. Crit Care 8:42–47
62. Chu VH, Woods CW, Miro JM et al (2008) Emergence of coagulase-negative staphylococci as a cause of native valve endocarditis. Clin Infect Dis 46:232–242
63. Castonguay MC, Burner KD, Edwards WD et al (2013) Surgical pathology of native valve endocarditis in 310 specimens from 287 patients (1985–2004). Cardiovasc Pathol 22:19–27
64. Caputo GM, Archer GL, Calderwood SB et al (1987) Native valve endocarditis due to coagulase-negative staphylococci. Clinical and microbiologic features. Am J Med 83:619–625
65. Chambers ST (2005) Diagnosis and management of staphylococcal infections of pacemakers and cardiac defibrillators. Intern Med J 35(Suppl 2):S63–S71
66. Athan E, Chu VH, Tattevin P et al (2012) Clinical characteristics and outcome of infective endocarditis involving implantable cardiac devices. JAMA 16:1727–1735
67. del Rio A, Anguera I, Miro JM et al (2003) Surgical treatment of pacemaker and defibrillator lead endocarditis: the impact of electrode lead extraction on outcome. Chest 124:1451–1459
68. Chua JD, Wilkoff BL, Lee I et al (2000) Diagnosis and management of infections involving implantable electrophysiologic cardiac devices. Ann Intern Med 133:604–608
69. Le KY, Sohail MR, Friedman PA et al (2012) Clinical features and outcomes of cardiovascular implantable electronic device infections due to staphylococcal species. Am J Cardiol 110:1143–1149
70. Baddour LM, Bettmann MA, Bolger AF et al (2003) Nonvalvular cardiovascular device-related infections. Circulation 108:2015–2031
71. Perera GB, Fujitani RM, Kubaska SM (2006) Aortic graft infection: update on management and treatment options. Vasc Endovascular Surg 40:1–10
72. Stewart AH, Eyers PS, Earnshaw JJ (2007) Prevention of infection in peripheral arterial reconstruction: a systematic review and meta-analysis. J Vasc Surg 46:148–155
73. Leroy O, Meybeck A, Sarraz-Bournet B et al (2012) Vascular graft infections. Curr Opin Infect Dis 25:154–158

74. Saleem BR, Meerwaldt R, Tielliu IFJ et al (2010) Conservative treatment of vascular prosthetic graft infection is associated with high mortality. Am J Surg 200:47–52
75. Trampuz A, Piper KE, Jacobson MJ et al (2007) Sonication of removed hip and knee prostheses for diagnosis of infection. N Engl J Med 357:654–663
76. Zimmerli W, Trampuz A, Ochsner PE (2004) Prosthetic-joint infections. N Engl J Med 351:1645–1654
77. Teterycz D, Ferry T, Lew D et al (2010) Outcome of orthopedic implant infections due to different staphylococci. Int J Infect Dis 14:913–918
78. Del Pozo JL, Patel R (2009) Infection associated with prosthetic joints. N Engl J Med 361:787–794
79. Kwee TC, Kwee RM, Abass A (2008) FDG-PTE for diagnosing joint infection: systematic review and metaanalysis. Eur J Nucl Med Mol Imaging 35:2122–2132
80. Atkins BL, Athanasou N, Deeks JJ et al (1998) Prospective evaluation of criteria for microbiological diagnosis of prosthetic-joint infection at revision arthroplasty. J Clin Microbiol 36:2932–2939
81. Vergidis P, Greenwood-Quaintance KE, Sanchez-Sotelo J et al (2011) Implant sonication for the diagnosis of prosthetic elbow infection. J Shoulder Elbow Surg 20:1275–1281
82. Maduka-Ezeh AN, Greenwood-Quaintance KE, Karau MJ et al (2012) Antimicrobial susceptibility and biofilm formation of *staphylococcus epidermidis* small colony variant associated with prosthetic joint infection. Diagn Microbiol Infect Dis 74:224–229
83. Singer J, Merz A, Frommelt L et al (2012) High rate of infection control with one-stage revision of septic knee prostheses excluding MRSA and MRSE. Clin Orthop Relat Res 470:1461–1471
84. Conen A, Walti LN, Merlo A et al (2008) Characteristics and treatment outcome of cerebrospinal fluid shunt-associated infections in adults: a retrospective analysis over an 11-year period. Clin Infect Dis 47:73–82
85. Tunkel AR, Hartman BJ, Kaplan SL et al (2004) Practice guidelines for the management of bacterial meningitis. Clin Infect Dis 39:1267–1284
86. Ratilal B, Costa J, Sampaio C (2006) Antibiotic prophylaxis for preventing meningitis in patients with basilar skull fractures. Cochrane Database Syst Rev 1, CD004884
87. Ritz R, Roser F, Morgalla M et al (2007) Do antibiotic-impregnated shunts in hydrocephalus therapy reduce the risk of infection? An observational study in 258 patients. BMC Infect Dis 7:38
88. Mangram AJ, Horan TC, Pearson ML et al (1999) Guideline for prevention of surgical site infection, 1999. Hospital Infection Control Practices Advisory Committee. Infect Control Hosp Epidemiol 20:250–278
89. Vas S, Oreopoulos DG (2001) Infections in patients undergoing peritoneal dialysis. Infect Dis Clin North Am 15:743–774
90. Baretti P, Montelli AC, Batalha JEN et al (2009) The role of virulence factors in the outcome of staphylococcal peritonitis in CAPD patients. BMC Infect Dis 9:212
91. Burke M, Hawley CM, Badve SV et al (2011) Relapsing and recurrent peritoneal dialysis-associated peritonitis: a multicenter registry study. Am J Kidney Dis 58:429–436
92. de Kaspar HM, Kreidl KO, Singh K et al (2004) Comparison of pre-operative conjunctival bacterial flora in patients undergoing glaucoma or cataract surgery. J Glaucoma 13: 507–509
93. Mollan SP, Gao A, Lockwood A et al (2007) Postcataract endophthalmitis: Incidence and microbial isolates in a United Kingdom region from 1996 through 2004. J Cataract Refract Surg 33:265–268
94. Hanscom TA (2004) Postoperative endophthalmitis. Clin Infect Dis 38:542–546
95. Haider SA, Hassett P, Bron AJ (2001) Intraocular vancomycin levels after intravitreal injection in post cataract extraction endophthalmitis. Retina 21:210–213
96. Carson CC (2003) Diagnosis, treatment and prevention of penile prosthesis infection. Int J Impot Res 15(Suppl 5):S139–S146
97. Darouiche RO (2004) Treatment of infections associated with surgical implants. N Engl J Med 350:1422–1429
98. Delgado S, Arroyo R, Jimenez E et al (2009) *Staphylococcus epidermidis* strains isolated from breast milk of women suffering infectious mastitis: potential virulence traits and resistance to antibiotics. BMC Microbiol 9:82
99. Del Pozo JL, Tran NV, Petty PM et al (2009) Pilot study of association of bacteria on breast implants with capsular contracture. J Clin Microbiol 47:1333–1337
100. Sampedro MF, Patel R (2007) Infections associated with long-term prosthetic devices. Infect Dis Clin North Am 21:785–819
101. Paluch-Oles J, Magrys A, Koziol-Montewka M et al (2011) The phenotypic and genetic biofilm formation characteristics of coagulase-negative staphylococci isolates in children with otitis media. Int J Pediatr Otorhinolaryngol 75:126–30
102. Whitfeld M, Gunasingam N, Leow LJ et al (2011) *Staphylococcus epidermidis*: a possible role in the pustules of rosacea. J Am Acad Dermatol 1:49–52

103. Davidson LE, Boucher HW (2009) Infections in immunocompromised patients. In: Crossley KB, Jefferson KK, Archer G, Fowler VG Jr (eds) Staphylococci in human disease, 2nd edn. Blackwell Publishing, Oxford
104. Sganga G, Spanu T, Bianco G et al (2012) Bacterial bloodstream infections in liver transplantation: etiologic agents and antimicrobial susceptibility profiles. Transplant Proc 44:1973–1976
105. Centers for Disease Control and Prevention. The national Healthcare Safety network (NHSN) Manual. Patient Safety Component. http://www.cdc.gov/nhsn/PDFs/pscManual/PSC-Manual-portfolio.pdf. Accessed 4 Apr 2013
106. Costa SF, Barone AA, Miceli MH et al (2006) Colonization and molecular epidemiology of coagulase-negative Staphylococcal bacteremia in cancer patients: a pilot study. Am J Infect Control 34:36–40
107. Kaufman D, Fairchild KD (2004) Clinical microbiology of bacterial and fungal sepsis in very-low-birth-weight infants. Clin Microbiol Rev 17:638–680
108. Stoll BJ, Hansen N, Fanaroff AA et al (2002) Late-onset sepsis in very low birth weight neonates: the experience of the NICHD Neonatal Research Network. Pediatrics 110(2002):285–291
109. Hira V, Kornelisse R, Slujter M et al (2013) Colonization dynamics of antibiotic-resistant coagulase-negative staphylococci in neonates. J Clin Microbiol 51:595–597
110. Dudeck MA, Horan TC, Peterson KD et al (2013) Centers for Disease Control and Prevention. National Healthcare Safety Network (NHSN) report, data summary for 2011, device-associated module. http://www.cdc.gov/nhsn.PDFs/dataStat/NHSN-Report-2011-Data-Summary.pdf. Accessed 4 April 2013
111. Brusselaers N, Monstrey S, Snoeij T et al (2010) Morbidity and mortality of bloodstream infections in patients with severe burn injury. Am J Crit Care 19:e81–e87
112. McCann MT, Gilmore BF, Gorman SP (2008) *Staphylococcus epidermidis* device-related infections: pathogenesis and clinical management. J Pharm Pharmacol 60:1551–1571
113. Nuryastuti T, van der Mei HC, Busscher HJ et al (2009) Effect of cinnamon oil on *ica*A expression and biofilm formation by *Staphylococcus epidermidis.* Appl Environ Microbiol 75:6850–6855
114. Hell E, Giske CG, Nelson A et al (2010) Human cathelicidin peptide LL37 inhibits both attachment capability and biofilm formation of *Staphylococcus epidermidis.* Lett Appl Microbiol 50:211–215
115. Wang X, Yao X, Zu Z et al (2009) Effect of berberine on *Staphylococcus epidermidis* biofilm formation. Int J Antimicrob Agents 34:60–66
116. Lonn-Stensrud J, Landin MA, Bennche T et al (2009) Furanones, potential agents for preventing *Staphylococcus epidermidis* biofilm infections? J Antimicrob Chemother 62: 309–16
117. Saader HS, Biedenbach DJ, Paukner S et al (2012) Antimicrobial activity of the investigational pleuromutilin compound BC-3781 tested against gram-positive organisms commonly associated with acute bacterial skin and skin structure infections. Antimicrob Agents Chemother 56:1619–1623
118. Francolini I, Donelli G (2010) Prevention and control of biofilm-based medical-device-related infections. FEMS Immunol Med Microbiol 59:227–238
119. Van Mellaert V, Shahrooei M, Hofmans D et al (2012) Immunoprophylaxis and immunotherapy of *Staphylococcus epidermidis* infections: challenges and prospects. Expert Rev Vaccines 11:319–334
120. Kaplan J (2009) Therapeutic potential of biofilm-dispersing enzymes. Int J Artif Organs 32:545–554
121. Phil M, Chavez de Paz LE, Schmidtchen A et al (2010) Effects of clinical isolates of *Pseudomonas aeurginosa* on *Staphylococcus epidermidis* biofilm formation. FEMS Immunol Med Microbiol 59:504–512
122. Harris LG, Bexfield A, Nigam Y et al (2009) Disruption of *Staphylococcus epidermidis* biofilms by medicinal maggot *Lucilia sericata* extcretions/secretions. Int J Artif Organs 32:555–564

Chapter 2

Staphylococcus epidermidis Pathogenesis

Michael Otto

Abstract

Staphylococcus epidermidis is the most frequently encountered member of the coagulase-negative staphylococci on human epithelial surfaces. It has emerged as an important nosocomial pathogen, especially in infections of indwelling medical devices. The mechanisms that *S. epidermidis* uses to survive during infection are in general of a passive nature, reflecting their possible origin in the commensal life of this bacterium. Most importantly, *S. epidermidis* excels in forming biofilms, sticky agglomerations that inhibit major host defense mechanisms. Furthermore, *S. epidermidis* produces a series of protective surface polymers and exoenzymes. Moreover, *S. epidermidis* has the capacity to secrete strongly cytolytic members of the phenol-soluble modulin (PSM) family, but PSMs in *S. epidermidis* overall appear to participate primarily in biofilm development. Finally, there is evidence for a virulence gene reservoir function of *S. epidermidis*, as it appears to have transferred important immune evasion and antibiotic resistance factors to *Staphylococcus aureus*. Conversely, *S. epidermidis* also has a beneficial role in balancing the microflora on human epithelial surfaces by controlling outgrowth of harmful bacteria such as in particular *S. aureus*. Recent research yielded detailed insight into key *S. epidermidis* virulence determinants and their regulation, in particular as far as biofilm formation is concerned, but we still have a serious lack of understanding of the in vivo relevance of many pathogenesis mechanisms and the factors that govern the commensal life of *S. epidermidis*.

Key words *Staphylococcus epidermidis*, Pathogenesis, Phenol-soluble modulins, Biofilm

1 Introduction

Research on staphylococcal virulence determinants has for the longest time focused mainly on the notorious human pathogen *Staphylococcus aureus*. However, in recent decades it has become increasingly clear that many members of the coagulase-negative staphylococci (CoNS) can cause disease [1]. *Staphylococcus epidermidis* is the most frequently encountered CoNS species on human skin and the premier cause of CoNS infections [2]. Characteristically, infections caused by *S. epidermidis* and other CoNS are chronic, which contrasts the potential of *S. aureus* to cause acute disease [3]. Furthermore, *S. epidermidis* infections are commonly nosocomial and only develop in predisposed patients, such as patients undergoing surgery or the immune compromised. *S. epidermidis* is

Paul D. Fey (ed.), *Staphylococcus Epidermidis: Methods and Protocols*, Methods in Molecular Biology, vol. 1106, DOI 10.1007/978-1-62703-736-5_2, © Springer Science+Business Media, LLC 2014

notorious in particular for causing infections on indwelling medical devices, in which pathogenesis usually involves the formation of surface-attached, sticky bacterial agglomerations called biofilms [4].

The increased interest in *S. epidermidis* as a pathogen has prompted the investigation of the molecular determinants underlying *S. epidermidis* disease pathogenesis. Given that *S. epidermidis* is mostly involved in biofilm-associated infections, most of these endeavors focused on factors promoting *S. epidermidis* biofilm development. Next to *Pseudomonas aeruginosa*, *S. epidermidis* now likely is the bacterium in which biofilm development is best understood. However, recent research also revealed that *S. epidermidis* produces a family of toxins, the phenol-soluble modulins (PSMs), with the potential to kill human red and white blood cells, potentially contributing to more aggressive forms of *S. epidermidis* disease [5, 6]. Moreover, it was recognized that *S. epidermidis* plays an important role as "reservoir" of genes that after horizontal gene transfer increase the pathogenic potential of *S. aureus*, such as antibiotic resistance determinants involved for example in methicillin resistance [7]. Finally, despite the increased awareness that *S. epidermidis* is an important pathogen, recent reports also addressed the role of *S. epidermidis* as a beneficial commensal that controls overgrowth of more aggressive human pathogens, most notably *S. aureus* [8]. Here, the most important molecular determinants underlying *S. epidermidis* pathogenesis, survival in its natural habitat, and competition with *S. aureus* are presented.

2 Biofilm Development

The formation of biofilms is the most important mechanism contributing to *S. epidermidis* infection. Additionally, *S. epidermidis* colonization of the human skin also arguably involves a biofilm-like state. Biofilm development commences with the attachment of cells to a surface, such as human tissue or the plastic surface of an indwelling medical device. However, indwelling medical devices soon are covered by human matrix proteins after insertion; and thus, direct attachment to plastic likely only plays a limited role during the establishment of *S. epidermidis* in vivo biofilms. Then, bacterial cells proliferate and produce a series of chemically different, secreted molecules that form the extracellular matrix, which holds biofilm cells together. Some of those factors may also originate from lysed cells, such as extracellular DNA (eDNA). Disruptive molecules create channels in the biofilm, which are important for nutrient accessibility in deeper biofilm layers and give the biofilm its characteristic structure, often described as mushroom-like. Finally, biofilm cells and clusters can detach in a passive way by sloughing or when triggered by high expression of biofilm-disruptive factors. This detachment process is of key importance for the dissemination of biofilm-associated infection [4, 9].

Attachment to abiotic surfaces such as the plastic surfaces of indwelling medical devices is determined mostly by the chemical composition of the bacterial cell surface and the resulting hydrophobicity. Some specific molecular determinants, such as the autolysin AtlE, have been implied in direct attachment of *S. epidermidis* to plastic [10], but the effect in those cases is likely indirect, inasmuch as AtlE is a strongly expressed protein whose absence considerably changes cell surface characteristics. Alteration of cell surface polymers such as teichoic acids or absence of eDNA may have similar effects [11, 12]. These effects are likely indirect due to differential binding of proteins to the cell surface.

In vivo, attachment occurs to host tissue or host matrix proteins. Like *S. aureus* and other staphylococci, *S. epidermidis* produces a variety of surface-located proteins that bind host proteins in a usually specific manner [13, 14]. Bacterial surface proteins with such capacities have been termed MSCRAMMS (microbial components recognizing adhesive matrix molecules) [15]. Most such proteins have a characteristic C-terminal sequence (LPXTG motif) that triggers sortase-mediated binding to the staphylococcal cell wall and is easily recognized by DNA sequence analysis [16]. According to genomic analyses, *S. epidermidis* has at least 14 MSCRAMMs with an LPXTG motif, which are thus presumably bound to the cell surface in a covalent fashion [13]. Many of those belong to the family of surface proteins containing serine/aspartate (SD) repeats, named Sdr proteins [17]. The SD repeats are supposed to stretch through the cell wall, exposing the binding domain at the cell wall surface. Other *S. epidermidis* surface proteins were named Ses, for *S. epidermidis* surface protein, and often yet lack a clearly attributed function [14]. Upon digestion with lysostaphin, which cleaves the interpeptide bridge of peptidoglycan, only some MSRCAMMs such as SdrF could be released [17]. Others, such as SdrG and SdrH, remained associated with the protoplast fraction, indicating ineffective sorting.

Several LPXTG motif-anchored surface proteins of *S. epidermidis* received special attention. The SD repeat family protein SdrG, also called Fbe for its fibrinogen-binding capacity, targets the thrombin cleavage site in the β chain of fibrinogen [18]. SdrG is necessary and sufficient for binding to fibrinogen-coated material. Its binding mechanism was described in detail using crystal structures and shown to follow what was called a "dock, lock, and latch" mechanism [19]. This mechanism presumably is used by a series of structurally related Gram-positive surface-binding proteins.

The accumulation-associated protein (Aap) forms zinc-dependent fibrils on the surface of *S. epidermidis* cells [20–22]. It is a major contributor to exopolysaccharide-independent biofilms of *S. epidermidis* (see below). In *S. aureus*, the Aap homologue SasG was later shown to have a comparable role [23], exemplifying how *S. aureus* biofilm research benefited from results achieved in *S. epidermidis*. Aap also has a role in attachment, inasmuch as it was shown to mediate adhesion to human corneocytes [24]. Corneocyte interaction is mediated by the terminal A domain of Aap, while the zinc-dependent fibril polymerization requires G5 domains and proteolytical removal of the terminal domains [22, 24, 25]. G5 domains interact with the biofilm exopolysaccharide polysaccharide intercellular adhesin (PIA, or poly-*N*-acetyl glucosamine, PNAG) to form a biofilm matrix network [22].

The already mentioned SdrF interacts with type I collagen [26] and was shown to contribute to the initiation of ventricular assist device driveline-related infections [27]. SdrF is only present in a subset of *S. epidermidis* strains [17]. Notably, SdrF had a more significant effect on the initiation of the tested device-related infection than other tested major *S. epidermidis* adhesive molecules (SdrG, PIA, and the lipase GehD), suggesting that SdrF-positive strains may have an advantage in colonizing indwelling medical devices [27].

The SesI protein, which is encoded on a prophage and limited to certain *S. epidermidis* strains, was suggested to be involved in *S. epidermidis* virulence, as based on a higher frequency in invasive *S. epidermidis* strains [28]. Recent research revealed a key role of the highly similar *S. aureus* SesI homologue, SasX, in nasal colonization and virulence, indicating that SesI may contribute to enhance these phenotypes also in *S. epidermidis* [29].

Some surface proteins that were implicated in attachment to plastic or tissue are anchored to the bacterial cell surface in a non-covalent fashion, likely using hydrophobic or electrostatic forces and possibly involving cell surface polymers such as teichoic acids [30]. The autolysin AtlE is involved in binding to plastic, although this is believed to be due to secondary effects, as outlined above. AtlE also mediates binding to vitronectin [10]. Another *S. epidermidis* bifunctional autolysin/adhesin, Aae, binds to fibrinogen, fibronectin, and vitronectin [31]. The giant ~1,000-kDa protein extracellular matrix-binding protein (Embp) mediates binding to fibronectin [32]. Finally, the secreted lipase GehD also appears to be involved in tissue binding, inasmuch as it mediates binding to collagen [33]. Obviously, this requires at least some fraction of the lipase to be non-covalently attached to the cell surface; and detection of GehD in lysostaphin cell extracts suggests that this is the case.

After attachment is accomplished, a series of chemically different molecules contribute to the formation of the extracellular biofilm matrix. Arguably the most important and certainly the molecule that has received most attention is PIA/PNAG. This linear beta-1,6-linked *N*-acetyl glucosamine polymer [34] is synthesized by the products of the *ica* gene cluster [35]. IcaA and IcaD form the *N*-acetyl glucosamine (GlcNAc) transferase that adds activated GlcNAc units to the growing PIA/PNAG chain [36]. Likely in connection with synthesis, the growing chain is exported by the membrane-spanning protein IcaC, although experimental evidence for the function of IcaC has not yet been produced [36]. After export, the surface-located IcaB enzyme partially deacetylates the GlcNAc homopolymer, introducing positive charges in the otherwise neutral molecule, which are crucial for surface location and the biological functions of PIA/PNAG [37].

Presence of the *ica* genes was linked to *S. epidermidis* invasiveness [38, 39], but this is controversial [40]. Furthermore, isolates lacking *ica* genes were found to be involved in biofilm-associated infection and may form in vitro biofilms [41, 42]. In these strains, proteins such as Aap represent primary constituents of the biofilm matrix [41]. In particular, the capacity of Aap to polymerize and form fibrils is deemed crucial for cell–cell interaction and matrix formation [21, 22]. As compared to *S. aureus*, *S. epidermidis* as a species appears to rely more frequently on exopolysaccharide (i.e., PIA/PNAG)-dependent than -independent biofilm formation. In vitro, PIA/PNAG-dependent biofilms have a rougher surface, while protein-dependent biofilms appear smoother [43]. Furthermore, PIA/PNAG-dependent biofilms seem to be more robust [41]. Finally, other nonprotein exopolymers such as teichoic acids and eDNA also contribute to the *S. epidermidis* biofilm matrix.

The characteristic structure of mature biofilms with mushroom-like shapes and channels is dependent on the production of cell–cell disruptive factors, which in *S. epidermidis* were shown to be PSMs [44]. These surfactant-like molecules presumably work by decreasing non-covalent adhesion between cells. In *S. epidermidis*, the β-type PSMs were shown to have such a role, but recent findings in *S. aureus* that attribute similar functions to all PSM classes [45] indicate that other *S. epidermidis* PSMs may also contribute to biofilm maturation.

Of primary importance for the dissemination of biofilm-associated infection, cells or cell clusters may detach from a biofilm to reach secondary infection sites. This may happen by mechanical forces under flow, such as present in a blood vessel, in a process often called sloughing. In addition, the bacteria can trigger detachment by PSM production. Production of β-type PSMs in *S. epidermidis* was shown to facilitate in vitro biofilm detachment and dissemination of biofilm-associated infection in vivo, a process that could be blocked by anti-PSM β antibodies [44]. Notably, the use of surfactant-like molecules to structure biofilms and cause detachment seems to be conserved among many bacteria, despite the fact that the surfactant molecules that contribute to these processes are chemically unrelated [46]. Especially in biofilms that depend on proteins or eDNA for biofilm formation, proteases or nucleases, respectively, were shown to detach biofilms in in vitro experiments [47, 48]. While these findings were achieved mostly in *S. aureus*, they can likely in principle be transferred to *S. epidermidis*. However, there is yet no evidence that such enzyme-based detachment mechanisms have in vivo relevance. In particular, the role of eDNA in in vivo biofilms is hypothetical. Presence of the potent DNAse I in human serum, which degrades in vitro biofilms [49], argues against such a role.

3 Interaction with Human Innate Host Defense

Professional phagocytes such as neutrophils limit staphylococcal proliferation after the bacteria breach through the epidermal protective layer. The bacteria are ingested by phagocytosis and then killed by antimicrobial peptides (AMPs) and proteins, as well as reactive oxygen species, in the neutrophil phagosome. In comparison to *S. aureus*, *S. epidermidis* is much more susceptible to elimination by neutrophils [50]. In *S. aureus*, cytolytic members of the PSM family have a primary role in eliminating neutrophils [51]. PSMs with cytolytic capacity usually belong to the PSMα type, which is distinguished from the β type by shorter size (~20–25 amino acids versus ~40 to 45 amino acids for β-type PSMs). Interestingly, *S. epidermidis* secretes strongly cytolytic PSMs, most notably the α-type PSMδ [5]. However, production of strongly cytolytic PSMs in *S. epidermidis* is very low, and the PSM pattern in *S. epidermidis* is shifted considerably to the non-cytolytic β-type PSMs [5, 52]. This difference may contribute significantly to the much lower aggressiveness of *S. epidermidis* as a pathogen compared to *S. aureus*.

The mechanisms that *S. epidermidis* uses to circumvent elimination by professional phagocytes are in general of a more "passive" nature. First, biofilm formation by itself contributes to resistance to phagocytosis [53]. Furthermore, *S. epidermidis* produces protective surface polymers that inhibit phagocytosis, such as PIA/PNAG [54] and poly-γ-glutamic acid [55], a polymer first described as component of the *Bacillus anthracis* capsule [56]. Moreover, secreted proteases such as *S. epidermidis* SepA may degrade bactericidal AMPs [57]. Accordingly, SepA has a significant impact on resistance to elimination by human neutrophils [5]. Finally, *S. epidermidis* can sense the presence of AMPs by a system called Aps, which upon activation up-regulates AMP-protective mechanisms, including the VraFG transporter, the D-alanylation of teichoic acids, and the lysylation of phospholipids [58]. The latter two mechanisms increase the positive charge of the cell surface, thereby increasing repulsion of the usually positively charged AMPs.

4 Toxins, Exoenzymes, and Other Virulence Determinants

In general, *S. epidermidis* does not produce toxins or other aggressive virulence factors. As already mentioned, cytolytic PSMs, although present, are not secreted at considerably toxic levels. It remains to be investigated whether differential production of these PSMs may be linked to *S. epidermidis* invasiveness. Furthermore, while the production of aggressive virulence determinants such as toxic shock syndrome toxin 1 (TSST-1), enterotoxin, and other superantigens has been described in CoNS [59], including *S. epidermidis*, this can be considered extremely rare. Recently, the first pathogenicity island harboring a gene for an enterotoxin has been described in *S. epidermidis* [60]; but genetic information on superantigens in *S. epidermidis* strains is not available. Notably, presence or production of such aggressive virulence determinants has not been correlated (yet) with *S. epidermidis* invasiveness.

Several secreted degradative enzymes of *S. epidermidis*, mostly proteases, were linked to virulence. The metalloprotease SepA degrades AMPs [57], and the glutamyl endopeptidase SspA (GluSE, Esp) degrades fibrinogen and complement factors [61]. All secreted *S. epidermidis* proteases—in addition to SspA and SepA and the cysteine protease SspB (Ecp) [61]—probably also degrade many other host factors and contribute to host tissue damage, likely to provide nutrients for the bacteria. They may also participate in the maturation process of other secreted enzymes [62]. The lipases GehC and GehD are believed to facilitate survival in fatty acid secretions [63, 64]. Furthermore, similar to *S. aureus*, *S. epidermidis* produces siderophores such as staphyloferrin to acquire essential iron from the host [65]. Finally, fatty acid-modifying enzyme (FAME) activity was described in *S. epidermidis* [66]. FAME inactivates host-derived fatty acids that are harmful for the bacteria, but the identity of FAME is unknown.

5 Regulatory Factors

Substantial efforts were undertaken to understand how *S. epidermidis* biofilm determinants are regulated. For example, a complex network of global regulators comprising SarA and different SarA homologues, SigB, LuxS, etc. [67–72] and environmental influences [73–75] regulate expression of *ica* and production of PIA/PNAG. This regulation may be dependent on the IcaR regulatory protein [73, 74] or directly target the promoter of the adjacent *icaADBC* biosynthetic operon. While regulation of PIA/PNAG is thus well analyzed on an in vitro molecular level, the biological significance of those regulatory influences for the in vivo situation is poorly understood. The Van Eldere group measured expression of key biofilm factors in vivo and found that genes for structural factors such as PIA/PNAG are expressed early during biofilm-associated infection [76]. However, the detailed regulatory processes and environmental influences governing the temporal expression of biofilm determinants in vivo remain largely uncharacterized.

Whereas the quorum-sensing regulator Agr does not impact expression of PIA/PNAG [77], it has a crucial impact on immune evasion and biofilm development, in part mediated via its strong influence on the production of PSMs, MSCRAMMs, and secreted degradative enzymes [62, 77–80]. In general, MSCRAMMS are known to be down-regulated by Agr, meaning that they are highly expressed during the beginning of an infection, while later on, with increased bacterial cell density, they become repressed. However, the in vivo temporal and spatial expression patterns of *S. epidermidis* Agr and MSCRAMMS await more detailed analysis. Interestingly, mutants in *agr* are frequently isolated from biofilm infections [79], indicating that functional Agr deficiency, which leads to more compact and extended biofilms, may be of benefit to the bacteria in such a situation. However, as confirmed by similar findings in *S. aureus*, *agr* mutants have lost the ability to disseminate and may be regarded as a dead end of infection [45, 79, 81].

6 In Vivo Analysis of Virulence

Much evidence on *S. epidermidis* virulence factors, especially those involved in biofilm formation, was achieved in vitro. However, the relevance of in vitro biofilm investigation for in vivo biofilm-associated infection situation is highly debatable [82]. Furthermore, the direct investigation of *S. epidermidis* virulence determinants is hampered by the difficulty to produce isogenic gene deletion mutants in *S. epidermidis*. Consequently, only some *S. epidermidis* virulence determinants have been investigated for their in vivo role using gene deletion mutants.

Most in vivo evidence is available for the role of PIA/PNAG, which proved essential for the development of *S. epidermidis* biofilm-associated infection in most animal models [83–86]. Of note, deacetylation of PIA/PNAG is crucial for in vivo infection [37]. Other *S. epidermidis* molecules for which roles in biofilm-associated infection were shown directly using isogenic gene deletion mutants in mouse, rat, or rabbit device-related infection with implanted catheters are AtlE [85], PSMβ peptides [44], and the quorum-sensing regulator Agr [79]. In addition, the Gao group investigated the role of several regulatory factors in a rat model of intravascular infection. Roles for the DNA-binding protein SarZ [72], the autoinducer II biosynthetic enzyme LuxS [68], the serine/threonine kinase Stk [87], the protease ClpP [88], and the general stress protein Ygs [89] in *S. epidermidis* virulence were detected using that model. However, these authors primarily used organ dissemination rather than biofilm formation on the catheters as readout; and thus, it remains unclear for those factors whether their established in vitro biofilm contribution is reflected in vivo, as the observed increased virulence potential may not have been biofilm related. Clearly, future efforts in *S. epidermidis* pathogenesis research must include the production of isogenic gene deletion mutants of additional determinants and their investigation using appropriate animal infection models, most notably of biofilm infection. Heterologous expression for example in *Lactococcus lactis* was used as an alternative method to investigate the role of *S. epidermidis* virulence determinants [27], but this method does not allow conclusions on the role of those factors in the natural strain background and should only be regarded as complementary.

7 Commensal Lifestyle

S. epidermidis is first and foremost a commensal organism on the human skin and mucous surfaces. Accordingly, most molecular determinants that were implicated in virulence have original functions in the nonpathogenic life of *S. epidermidis* but may rise to additional benefit during infection [90]. The event of breaching the epidermal protective barrier may be seen as accident rather than a pathogenesis program used by *S. epidermidis*. Accordingly, only a very limited number of factors could be linked by molecular epidemiology to *S. epidermidis* invasiveness, including *ica* [39] and the insertion element IS256 [91, 92], which is believed to contribute to genetic adaptation [93, 94]. Examples of original functions of virulence-associated molecules in the commensal state include the surfactant function of the PSM cytolysins [95], the osmotolerance function of PGA [55], the cell division function of bifunctional autolysins/adhesins such as AtlE [96], and MSCRAMM-mediated binding to skin cells. Unfortunately, due to the lack of appropriate animal skin colonization models, our understanding of the factors that facilitate survival of *S. epidermidis* on the human skin is largely hypothetical. Very rarely, human colonization models were used [97].

8 Probiotic Function of *S. epidermidis*

Competition between the relatively innocuous *S. epidermidis* and the more aggressive pathogen *S. aureus* may have a considerable role in balancing the human epithelial microflora and preventing the overgrowth of bacteria that are harmful to the host. Interference based on Agr subgroups appears to be in general in favor of *S. epidermidis* [98], but there is no evidence linking Agr interference to the superiority of *S. epidermidis* as a colonizer. Recently, Iwase et al. could demonstrate that *S. epidermidis* strains expressing the protease SepA outcompete *S. aureus* during nasal colonization, while strains not producing SepA do not have that capacity [8]. The authors explained this by in vitro results indicating an impact of SepA on *S. aureus* biofilms, which are often protein dependent, but whether this mechanism accounts for the in vivo observations is not entirely clear [99]. Nevertheless, this study showed that *S. epidermidis* might have an important probiotic function by limiting *S. aureus* nasal colonization, which is key to subsequent infection with that pathogen [100].

9 Virulence Gene Reservoir for *S. aureus*

S. epidermidis may be in close contact with *S. aureus* in its natural habitat and transfer genetic information to that pathogen. Interestingly, this transfer appears to be unidirectional, inasmuch as aggressive virulence determinants of *S. aureus* are not found in *S. epidermidis* [7]. Possibly, clustered regularly interspaced short palindromic repeat (CRISPR)-mediated restriction of DNA uptake plays a role in that process [101]. In contrast, there is evidence indicating that factors facilitating *S. aureus* pathogenesis originated from *S. epidermidis*. These include several types of methicillin resistance (SCC*mec*) elements, most notably the low-fitness cost type IV SCC*mec* element that is characteristic of community-associated methicillin-resistant *S. aureus* (CA-MRSA) [102]. Furthermore, the USA300 CA-MRSA clone very likely acquired the arginine catabolic mobile genetic element (ACME) from *S. epidermidis*, where it is widespread [103]. The polyamine-detoxifying *speG* gene on this element was recently linked to survival and virulence of strain USA300 [104]. Finally, the prophage harboring the *sesI/sasX* gene is found in *S. epidermidis* and may thus also originate from CoNS [29, 105].

10 Outlook and Concluding Remarks

While research on *S. epidermidis* pathogenesis has progressed significantly over the last two decades, key problems remain. First, with few exceptions, it has remained almost impossible to manipulate clinical *S. epidermidis* strains due to inefficient transformation methods. Second, there should be a stronger focus on in vivo investigation, especially in the biofilm field, where in vitro results often only have limited bearing for in vivo infection. Third, models of biofilm-associated infection should be optimized to better reflect the clinical situation. Fourth, the molecular factors governing survival of *S. epidermidis* on the skin are still almost completely unknown, requiring the setup and use of appropriate skin colonization models. Considerable technical efforts are therefore necessary to further our understanding of how *S. epidermidis* colonizes, interacts with other bacteria, and causes disease.

Acknowledgements

This work was supported by the Intramural Research Program of the National Institute of Allergy and Infectious Diseases (NIAID), US National Institutes of Health.

References

1. Rogers KL, Fey PD, Rupp ME (2009) Coagulase-negative staphylococcal infections. Infect Dis Clin North Am 23:73–98
2. Vuong C, Otto M (2002) *Staphylococcus epidermidis* infections. Microbes Infect 4: 481–489
3. Lowy FD (1998) *Staphylococcus aureus* infections. N Engl J Med 339:520–532
4. Otto M (2008) Staphylococcal biofilms. Curr Top Microbiol Immunol 322:207–228
5. Cheung GY, Rigby K, Wang R et al (2010) *Staphylococcus epidermidis* strategies to avoid killing by human neutrophils. PLoS Pathog 6:e1001133
6. Mehlin C, Headley CM, Klebanoff SJ (1999) An inflammatory polypeptide complex from *Staphylococcus epidermidis*: isolation and characterization. J Exp Med 189:907–918
7. Otto M (2013) Coagulase-negative staphylococci as reservoirs of genes facilitating MRSA infection: Staphylococcal commensal species such as *Staphylococcus epidermidis* are being recognized as important sources of genes promoting MRSA colonization and virulence. Bioessays 35:4–11
8. Iwase T, Uehara Y, Shinji H et al (2010) *Staphylococcus epidermidis* Esp inhibits *Staphylococcus aureus* biofilm formation and nasal colonization. Nature 465:346–349
9. O'Toole G, Kaplan HB, Kolter R (2000) Biofilm formation as microbial development. Annu Rev Microbiol 54:49–79
10. Heilmann C, Hussain M, Peters G et al (1997) Evidence for autolysin-mediated primary attachment of *Staphylococcus epidermidis* to a polystyrene surface. Mol Microbiol 24:1013–1024
11. Holland LM, Conlon B, O'Gara JP (2011) Mutation of *tagO* reveals an essential role for wall teichoic acids in *Staphylococcus epidermidis* biofilm development. Microbiology 157:408–418
12. Christner M, Heinze C, Busch M et al (2012) *sarA* negatively regulates *Staphylococcus epidermidis* biofilm formation by modulating expression of 1 MDa extracellular matrix binding protein and autolysis-dependent release of eDNA. Mol Microbiol 86:394–410
13. Gill SR, Fouts DE, Archer GL et al (2005) Insights on evolution of virulence and resistance from the complete genome analysis of an early methicillin-resistant *Staphylococcus aureus* strain and a biofilm-producing methicillin-resistant *Staphylococcus epidermidis* strain. J Bacteriol 187:2426–2438
14. Bowden MG, Chen W, Singvall J et al (2005) Identification and preliminary characterization of cell-wall-anchored proteins of *Staphylococcus epidermidis*. Microbiology 151:1453–1464
15. Patti JM, Allen BL, McGavin MJ et al (1994) MSCRAMM-mediated adherence of microorganisms to host tissues. Annu Rev Microbiol 48:585–617
16. Mazmanian SK, Liu G, Ton-That H et al (1999) *Staphylococcus aureus* sortase, an enzyme that anchors surface proteins to the cell wall. Science 285:760–763
17. McCrea KW, Hartford O, Davis S et al (2000) The serine-aspartate repeat (Sdr) protein family in *Staphylococcus epidermidis*. Microbiology 146:1535–1546
18. Davis SL, Gurusiddappa S, McCrea KW et al (2001) SdrG, a fibrinogen-binding bacterial adhesin of the microbial surface components recognizing adhesive matrix molecules subfamily from *Staphylococcus epidermidis*, targets the thrombin cleavage site in the Bbeta chain. J Biol Chem 276:27799–27805
19. Ponnuraj K, Bowden MG, Davis S et al (2003) A "dock, lock, and latch" structural model for a staphylococcal adhesin binding to fibrinogen. Cell 115:217–228
20. Hussain M, Herrmann M, von Eiff C et al (1997) A 140-kilodalton extracellular protein is essential for the accumulation of *Staphylococcus epidermidis* strains on surfaces. Infect Immun 65:519–524
21. Banner MA, Cunniffe JG, Macintosh RL et al (2007) Localized tufts of fibrils on *Staphylococcus epidermidis* NCTC 11047 are comprised of the accumulation-associated protein. J Bacteriol 189:2793–2804
22. Conrady DG, Brescia CC, Horii K, Weiss AA, Hassett DJ, Herr AB (2008) A zinc-dependent adhesion module is responsible for

intercellular adhesion in staphylococcal biofilms. Proc Natl Acad Sci U S A 105: 19456–19461
23. Corrigan RM, Rigby D, Handley P et al (2007) The role of *Staphylococcus aureus* surface protein SasG in adherence and biofilm formation. Microbiology 153:2435–2446
24. Macintosh RL, Brittan JL, Bhattacharya R et al (2009) The terminal A domain of the fibrillar accumulation-associated protein (Aap) of *Staphylococcus epidermidis* mediates adhesion to human corneocytes. J Bacteriol 191:7007–7016
25. Rohde H, Burdelski C, Bartscht K et al (2005) Induction of *Staphylococcus epidermidis* biofilm formation via proteolytic processing of the accumulation-associated protein by staphylococcal and host proteases. Mol Microbiol 55:1883–1895
26. Arrecubieta C, Lee MH, Macey A et al (2007) SdrF, a *Staphylococcus epidermidis* surface protein, binds type I collagen. J Biol Chem 282:18767–18776
27. Arrecubieta C, Toba FA, von Bayern M et al (2009) SdrF, a *Staphylococcus epidermidis* surface protein, contributes to the initiation of ventricular assist device driveline-related infections. PLoS Pathog 5:e1000411
28. Soderquist B, Andersson M, Nilsson M et al (2009) *Staphylococcus epidermidis* surface protein I (SesI): a marker of the invasive capacity of *S. epidermidis*? J Med Microbiol 58:1395–1397
29. Li M, Du X, Villaruz AE et al (2012) MRSA epidemic linked to a quickly spreading colonization and virulence determinant. Nat Med 18:816–819
30. Heilmann C (2011) Adhesion mechanisms of staphylococci. Adv Exp Med Biol 715:105–123
31. Heilmann C, Thumm G, Chhatwal GS et al (2003) Identification and characterization of a novel autolysin (Aae) with adhesive properties from *Staphylococcus epidermidis*. Microbiology 149:2769–2778
32. Christner M, Franke GC, Schommer NN et al (2010) The giant extracellular matrix-binding protein of *Staphylococcus epidermidis* mediates biofilm accumulation and attachment to fibronectin. Mol Microbiol 75:187–207
33. Bowden MG, Visai L, Longshaw CM et al (2002) Is the GehD lipase from *Staphylococcus epidermidis* a collagen binding adhesin? J Biol Chem 277:43017–43023
34. Mack D, Fischer W, Krokotsch A et al (1996) The intercellular adhesin involved in biofilm accumulation of *Staphylococcus epidermidis* is a linear beta-1,6-linked glucosaminoglycan: purification and structural analysis. J Bacteriol 178:175–183
35. Heilmann C, Schweitzer O, Gerke C et al (1996) Molecular basis of intercellular adhesion in the biofilm-forming *Staphylococcus epidermidis*. Mol Microbiol 20:1083–1091
36. Gerke C, Kraft A, Sussmuth R et al (1998) Characterization of the N-acetylglucosaminyltransferase activity involved in the biosynthesis of the *Staphylococcus epidermidis* polysaccharide intercellular adhesin. J Biol Chem 273:18586–18593
37. Vuong C, Kocianova S, Voyich JM et al (2004) A crucial role for exopolysaccharide modification in bacterial biofilm formation, immune evasion, and virulence. J Biol Chem 279:54881–54886
38. Li H, Xu L, Wang J et al (2005) Conversion of *Staphylococcus epidermidis* strains from commensal to invasive by expression of the *ica* locus encoding production of biofilm exopolysaccharide. Infect Immun 73:3188–3191
39. Galdbart JO, Allignet J, Tung HS et al (2000) Screening for *Staphylococcus epidermidis* markers discriminating between skin-flora strains and those responsible for infections of joint prostheses. J Infect Dis 182:351–355
40. Rohde H, Kalitzky M, Kroger N et al (2004) Detection of virulence-associated genes not useful for discriminating between invasive and commensal *Staphylococcus epidermidis* strains from a bone marrow transplant unit. J Clin Microbiol 42:5614–5619
41. Rohde H, Burandt EC, Siemssen N et al (2007) Polysaccharide intercellular adhesin or protein factors in biofilm accumulation of *Staphylococcus epidermidis* and *Staphylococcus aureus* isolated from prosthetic hip and knee joint infections. Biomaterials 28:1711–1720
42. Kogan G, Sadovskaya I, Chaignon P et al (2006) Biofilms of clinical strains of *Staphylococcus* that do not contain polysaccharide intercellular adhesin. FEMS Microbiol Lett 255:11–16
43. Schommer NN, Christner M, Hentschke M et al (2011) *Staphylococcus epidermidis* uses distinct mechanisms of biofilm formation to interfere with phagocytosis and activation of mouse macrophage-like cells 774A.1. Infect Immun 79:2267–2276
44. Wang R, Khan BA, Cheung GY et al (2011) *Staphylococcus epidermidis* surfactant peptides promote biofilm maturation and dissemination of biofilm-associated infection in mice. J Clin Invest 121:238–248
45. Periasamy S, Joo HS, Duong AC et al (2012) How *Staphylococcus aureus* biofilms develop their characteristic structure. Proc Natl Acad Sci U S A 109:1281–1286
46. Otto M (2013) Staphylococcal infections: mechanisms of biofilm maturation and

detachment as critical determinants of pathogenicity. Annu Rev Med 64:175–188
47. Kiedrowski MR, Kavanaugh JS, Malone CL et al (2011) Nuclease modulates biofilm formation in community-associated methicillin-resistant *Staphylococcus aureus*. PLoS One 6:e26714
48. Boles BR, Horswill AR (2008) Agr-mediated dispersal of *Staphylococcus aureus* biofilms. PLoS Pathog 4:e1000052
49. Whitchurch CB, Tolker-Nielsen T et al (2002) Extracellular DNA required for bacterial biofilm formation. Science 295:1487
50. Rogers DE, Tompsett R (1952) The survival of staphylococci within human leukocytes. J Exp Med 95:209–230
51. Wang R, Braughton KR, Kretschmer D et al (2007) Identification of novel cytolytic peptides as key virulence determinants for community-associated MRSA. Nat Med 13:1510–1514
52. Yao Y, Sturdevant DE, Otto M (2005) Genomewide analysis of gene expression in *Staphylococcus epidermidis* biofilms: insights into the pathophysiology of *S. epidermidis* biofilms and the role of phenol-soluble modulins in formation of biofilms. J Infect Dis 191:289–298
53. Cerca N, Jefferson KK, Oliveira R et al (2006) Comparative antibody-mediated phagocytosis of *Staphylococcus epidermidis* cells grown in a biofilm or in the planktonic state. Infect Immun 74:4849–4855
54. Vuong C, Voyich JM, Fischer ER et al (2004) Polysaccharide intercellular adhesin (PIA) protects *Staphylococcus epidermidis* against major components of the human innate immune system. Cell Microbiol 6:269–275
55. Kocianova S, Vuong C, Yao Y et al (2005) Key role of poly-gamma-DL-glutamic acid in immune evasion and virulence of *Staphylococcus epidermidis*. J Clin Invest 115:688–694
56. Hanby WE, Rydon HN (1946) The capsular substance of *Bacillus anthracis*. Biochemistry 40:297–307
57. Lai Y, Villaruz AE, Li M, Cha DJ et al (2007) The human anionic antimicrobial peptide dermcidin induces proteolytic defence mechanisms in staphylococci. Mol Microbiol 63:497–506
58. Li M, Lai Y, Villaruz AE et al (2007) Gram-positive three-component antimicrobial peptide-sensing system. Proc Natl Acad Sci U S A 104:9469–9474
59. Otto M (2004) Virulence factors of the coagulase-negative staphylococci. Front Biosci 9:841–863
60. Madhusoodanan J, Seo KS, Remortel B et al (2011) An enterotoxin-bearing pathogenicity island in *Staphylococcus epidermidis*. J Bacteriol 193:1854–1862
61. Dubin G, Chmiel D, Mak P et al (2001) Molecular cloning and biochemical characterisation of proteases from *Staphylococcus epidermidis*. Biol Chem 382:1575–1582
62. Vuong C, Gotz F, Otto M (2000) Construction and characterization of an *agr* deletion mutant of *Staphylococcus epidermidis*. Infect Immun 68:1048–1053
63. Farrell AM, Foster TJ, Holland KT (1993) Molecular analysis and expression of the lipase of *Staphylococcus epidermidis*. J Gen Microbiol 139:267–277
64. Longshaw CM, Farrell AM, Wright JD et al (2000) Identification of a second lipase gene, gehD, in *Staphylococcus epidermidis*: comparison of sequence with those of other staphylococcal lipases. Microbiology 146:1419–1427
65. Lindsay JA, Riley TV, Mee BJ (1994) Production of siderophore by coagulase-negative staphylococci and its relation to virulence. Eur J Clin Microbiol Infect Dis 13:1063–1066
66. Chamberlain NR, Brueggemann SA (1997) Characterisation and expression of fatty acid modifying enzyme produced by *Staphylococcus epidermidis*. J Med Microbiol 46:693–697
67. Li M, Villaruz AE, Vadyvaloo V et al (2008) AI-2-dependent gene regulation in *Staphylococcus epidermidis*. BMC Microbiol 8:4
68. Xu L, Li H, Vuong C et al (2006) Role of the *luxS* quorum-sensing system in biofilm formation and virulence of *Staphylococcus epidermidis*. Infect Immun 74:488–496
69. Handke LD, Slater SR, Conlon KM et al (2007) SigmaB and SarA independently regulate polysaccharide intercellular adhesin production in *Staphylococcus epidermidis*. Can J Microbiol 53:82–91
70. Kies S, Otto M, Vuong C et al (2001) Identification of the *sigB* operon in *Staphylococcus epidermidis*: construction and characterization of a *sigB* deletion mutant. Infect Immun 69:7933–7936
71. Knobloch JK, Jager S, Horstkotte MA et al (2004) RsbU-dependent regulation of *Staphylococcus epidermidis* biofilm formation is mediated via the alternative sigma factor sigmaB by repression of the negative regulator gene *icaR*. Infect Immun 72:3838–3848
72. Wang L, Li M, Dong D et al (2008) SarZ is a key regulator of biofilm formation and virulence in *Staphylococcus epidermidis*. J Infect Dis 197:1254–1262
73. Conlon KM, Humphreys H, O'Gara JP (2002) Regulation of *icaR* gene expression in *Staphylococcus epidermidis*. FEMS Microbiol Lett 216:171–177

74. Conlon KM, Humphreys H, O'Gara JP (2002) *icaR* encodes a transcriptional repressor involved in environmental regulation of *ica* operon expression and biofilm formation in *Staphylococcus epidermidis*. J Bacteriol 184:4400–4408
75. Knobloch JK, Horstkotte MA, Rohde H et al (2002) Alcoholic ingredients in skin disinfectants increase biofilm expression of *Staphylococcus epidermidis*. J Antimicrob Chemother 49:683–687
76. Vandecasteele SJ, Peetermans WE, Merckx R et al (2003) Expression of biofilm-associated genes in *Staphylococcus epidermidis* during in vitro and in vivo foreign body infections. J Infect Dis 188:730–737
77. Vuong C, Gerke C, Somerville GA et al (2003) Quorum-sensing control of biofilm factors in *Staphylococcus epidermidis*. J Infect Dis 188:706–718
78. Vuong C, Durr M, Carmody AB et al (2004) Regulated expression of pathogen-associated molecular pattern molecules in *Staphylococcus epidermidis*: quorum-sensing determines proinflammatory capacity and production of phenol-soluble modulins. Cell Microbiol 6:753–759
79. Vuong C, Kocianova S, Yao Y et al (2004) Increased colonization of indwelling medical devices by quorum-sensing mutants of *Staphylococcus epidermidis* in vivo. J Infect Dis 190:1498–1505
80. Yao Y, Vuong C, Kocianova S et al (2006) characterization of the *Staphylococcus epidermidis* accessory-gene regulator response: quorum-sensing regulation of resistance to human innate host defense. J Infect Dis 193:841–848
81. Shopsin B, Eaton C, Wasserman GA et al (2010) Mutations in *agr* do not persist in natural populations of methicillin-resistant *Staphylococcus aureus*. J Infect Dis 202: 1593–1599
82. Joo HS, Otto M (2012) Molecular basis of in vivo biofilm formation by bacterial pathogens. Chem Biol 19:1503–1513
83. Rupp ME, Ulphani JS, Fey PD et al (1999) Characterization of the importance of polysaccharide intercellular adhesin/hemagglutinin of *Staphylococcus epidermidis* in the pathogenesis of biomaterial-based infection in a mouse foreign body infection model. Infect Immun 67:2627–2632
84. Rupp ME, Ulphani JS, Fey PD et al (1999) Characterization of *Staphylococcus epidermidis* polysaccharide intercellular adhesin/hemagglutinin in the pathogenesis of intravascular catheter-associated infection in a rat model. Infect Immun 67:2656–2659
85. Rupp ME, Fey PD, Heilmann C et al (2001) Characterization of the importance of *Staphylococcus epidermidis* autolysin and polysaccharide intercellular adhesin in the pathogenesis of intravascular catheter-associated infection in a rat model. J Infect Dis 183:1038–1042
86. Begun J, Gaiani JM, Rohde H et al (2007) Staphylococcal biofilm exopolysaccharide protects against *Caenorhabditis elegans* immune defenses. PLoS Pathog 3:e57
87. Liu Q, Fan J, Niu C et al (2011) The eukaryotic-type serine/threonine protein kinase Stk is required for biofilm formation and virulence in *Staphylococcus epidermidis*. PLoS ONE 6:e25380
88. Wang C, Fan J, Niu C et al (2010) Role of *spx* in biofilm formation of *Staphylococcus epidermidis*. FEMS Immunol Med Microbiol 59:52–160
89. Wang X, Niu C, Sun G et al (2011) *ygs* is a novel gene that influences biofilm formation and the general stress response of *Staphylococcus epidermidis*. Infect Immun 79:1007–1015
90. Otto M (2009) *Staphylococcus epidermidis*—the 'accidental' pathogen. Nat Rev Microbiol 7:555–567
91. Gu J, Li H, Li M et al (2005) Bacterial insertion sequence IS256 as a potential molecular marker to discriminate invasive strains from commensal strains of *Staphylococcus epidermidis*. J Hosp Infect 61:342–348
92. Kozitskaya S, Cho SH, Dietrich K et al (2004) The bacterial insertion sequence element IS256 occurs preferentially in nosocomial *Staphylococcus epidermidis* isolates: association with biofilm formation and resistance to aminoglycosides. Infect Immun 72:1210–1215
93. Schoenfelder SM, Lange C, Eckart M et al (2010) Success through diversity—how *Staphylococcus epidermidis* establishes as a nosocomial pathogen. Int J Med Microbiol 300:380–386
94. Ziebuhr W, Krimmer V, Rachid S et al (1999) A novel mechanism of phase variation of virulence in *Staphylococcus epidermidis*: evidence for control of the polysaccharide intercellular adhesin synthesis by alternating insertion and excision of the insertion sequence element IS256. Mol Microbiol 32:345–356
95. Periasamy S, Chatterjee SS, Cheung GY et al (2012) Phenol-soluble modulins in staphylococci: What are they originally for? Commun Integr Biol 5:275–277
96. Biswas R, Voggu L, Simon UK et al (2006) Activity of the major staphylococcal autolysin Atl. FEMS Microbiol Lett 259:260–268
97. Rogers KL, Rupp ME, Fey PD (2008) The presence of *icaADBC* is detrimental to the

colonization of human skin by *Staphylococcus epidermidis*. Appl Environ Microbiol 74:6155–6157

98. Otto M, Sussmuth R, Vuong C et al (1999) Inhibition of virulence factor expression in *Staphylococcus aureus* by the *Staphylococcus epidermidis agr* pheromone and derivatives. FEBS Lett 450:257–262
99. Krismer B, Peschel A (2011) Does *Staphylococcus aureus* nasal colonization involve biofilm formation? Future Microbiol 6:489–493
100. von Eiff C, Becker K, Machka K et al (2001) Nasal carriage as a source of *Staphylococcus aureus* bacteremia. Study Group. N Engl J Med 344:11–16
101. Marraffini LA, Sontheimer EJ (2010) CRISPR interference: RNA-directed adaptive immunity in bacteria and archaea. Nat Rev Genet 11:181–190
102. Daum RS, Ito T, Hiramatsu K et al (2002) A novel methicillin-resistance cassette in community-acquired methicillin-resistant *Staphylococcus aureus* isolates of diverse genetic backgrounds. J Infect Dis 186: 1344–1347
103. Diep BA, Gill SR, Chang RF et al (2006) Complete genome sequence of USA300, an epidemic clone of community-acquired meticillin-resistant *Staphylococcus aureus*. Lancet 367:731–739
104. Joshi GS, Spontak JS, Klapper DG et al (2011) Arginine catabolic mobile element encoded speG abrogates the unique hypersensitivity of *Staphylococcus aureus* to exogenous polyamines. Mol Microbiol 82:9–20
105. Holden MT, Lindsay JA, Corton C et al (2010) Genome sequence of a recently emerged, highly transmissible, multi-antibiotic- and antiseptic-resistant variant of methicillin-resistant *Staphylococcus aureus*, sequence type 239 (TW). J Bacteriol 192: 888–892

Chapter 3

Identification of *Staphylococcus epidermidis* in the Clinical Microbiology Laboratory by Molecular Methods

Amity L. Roberts

Abstract

Biochemical assays for the phenotypic identification of coagulase-negative staphylococci in the clinical microbiology laboratory have been well described in previous publications (Becker and Von Eiff Manual of Clinical Microbiology, ASM Press, Washington, pp. 308–330, 2011; Kloos and Wolfshohl J Clin Microbiol 16:509–516, 1982). This discussion focuses on identification of *Staphylococcus epidermidis* through molecular and proteomic methods. Molecular assays have been shown to be more discriminatory between the coagulase-negative staphylococcal species than are phenotypic assays (Zadoks and Watts Vet Microbiol 134:20–28, 2009; Sheraba et al. BMC Res Notes 3:278, 2010; Patteet et al. Eur J Clin Microbiol Infect Dis 31:747–751, 2012). The molecular and proteomic methods that have shown the greatest utilization potential within the clinical laboratory are as follows: PCR amplification and sequencing of discriminatory genes, real-time polymerase chain reaction with species-specific probes in conjunction with a melt-curve analysis, fluorescence in situ hybridization, and matrix-assisted laser desorption-ionization time-of-flight mass spectrometry.

Key words Real-time PCR, MALDI-TOF-MS, PCR, Molecular, Proteomic, Sequencing, Fluorescence in situ hybridization

1 Introduction

While many clinical microbiology laboratories predominantly utilize biochemical assays to identify staphylococci to the species level, a number of clinical laboratories have begun implementing molecular or proteomic based methods. The biochemical assays predominantly utilized within the clinical microbiology laboratory, for identifying Gram-positive bacteria, are the API 20 Staph system (bioMérieux), ID 32 Staph API system (bioMérieux), Staph-Zym system (Rosco), Vitek Gram-positive system (bioMérieux), and Microscan WalkAway-96 POS Combo type 28 or 33 panels (Siemens Healthcare Diagnostics) [1–4]. Rapid differentiation of coagulase-negative *Staphylococcus* (CONS) from *Staphylococcus aureus* is facilitated by a negative staph-latex agglutination assay

Paul D. Fey (ed.), *Staphylococcus Epidermidis: Methods and Protocols*, Methods in Molecular Biology, vol. 1106,
DOI 10.1007/978-1-62703-736-5_3,

and a negative slide or a tube serum coagulase test. More stringent biochemical tests are necessary to identify CONS to the species level. These biochemical tests have been well defined in reviews by Kloos and Versalovic [5, 6].

The advent of PCR, real-time polymerase chain reaction (real-time PCR), next-generation sequencing, and matrix-assisted laser desorption-ionization time-of-flight mass spectrometry (MALDI-TOF-MS) has revolutionized microbial identification. These assays are now affordable, more rapid, and highly specific. Original PCR amplification assays of certain conserved genes focused on species-specific DNA banding patterns; however, sequencing of the specific gene and comparison of the sequence to that of known sequences within a database such as NCBI Basic Local Alignment Search Tool (BLAST) is a more specific method for identification to the species level [7–10]. The methods discussed within this text detail how to perform PCR amplification for sequencing, real-time PCR, Fluorescence In Situ Hybridization (FISH), and MALDI-TOF-MS specific to *S. epidermidis.*

2 Materials

2.1 PCR Amplification Components

1. Probes and primers: Using molecular grade H_2O, adjust lyophilized material to a 100 pmol/μl stock concentration and store at −20 °C. A working stock of the desired concentration (*see* Table 1) should be prepared when ready to perform the assay.
2. *Taq* DNA polymerase and 10× *Taq* buffer; store at −20 °C.
3. Tris–Borate–EDTA (TBE): Utilize 0.5× TBE for both the preparation of and running of the DNA non-denaturing 2 % agarose gel. Make 10× TBE by mixing Tris base 108 g, boric acid 55 g, and 4.65 g EDTA (or 40 ml 0.5 M EDTA; pH 8.0) with 900 ml of distilled H_2O (dH_2O) on a magnetic stir plate. Adjust the pH to 8.3, while stirring, using 1 N NaOH and then adding dH_2O to a total volume of 1 L. Dilute 10× TBE to 0.5× TBE (50 ml 10× TBE into 950 ml dH_2O). Store solution at room temperature in a sealed container.
4. 2 % agarose gel for resolution of fragments 0.2–1 Kb: Weigh out the agarose powder (1 g for a 50 ml gel or 2 g for a 100 ml gel mold). Add agarose to a 250 ml conical flask and fill with 50 or 100 ml of 0.5× TBE buffer. Microwave for 1.5–3 min, gently swirling the flask every 30 s. Use a heat-resistant hand protector to remove flask from the microwave; the agarose should be dissolved and the solution clear. Allow the agarose solution to cool to ~60 °C. As agarose solution is cooling, set up a horizontal electrophoresis chamber that will be referred to as the gel rig. Seal the ends of the UV transparent gel tray with either casting gates or wide laboratory tape. Position the

Table 1
Primers and amplification conditions

Target	Primers for amplification	PCR amplification conditions	Source/s	Product size
16S rRNA gene	SSU-bact-27f 5′-AGAGTTTGATCM TGGCTCAG-3′ SSU-bact-907r 5′-CCGTCAATTC MTTTRAGTTT-3′ Epsilon F 5′-AAGAGTTTGATCCT GGCTCCAG-3′ 1510R 5′-GGTTACCTTGTTACGACTT-3′	94 °C 5 min 30 cycles: –94 °C 45 s –53 °C 60 s –72 °C 90 s 72 °C 10 min	[9, 11, 14, 15, 23]	
16S–23S internal transcribed spacer sequence	G1(F):: 5′-GAAGTCGTAACAAGG-3′ L1(R) 5′-CAAGGCATCCACCGT-3′	94 °C 4 min 25 cycles: –94 °C 1 min –Ramp to 55 °C 2 min –55 °C 7 min –Ramp to 72 °C 2 min –72 °C 2 min Final extension 72 °C 7 min	[7, 13, 20]	
tuf	TStaG422 (F) 5′-GGCCGTGTTGAACGT GGTCAAATCA-3′ TStag765 (R) 5′-TIACCATTTCAGTAC CTTCTGGTAA-3′ *tuf*-F 5′-GCCAGTTGAGGACGTATTCT-3′ *tuf*-R 5′-CCATTTCAGTACCTTCTGGTAA-3′	96 °C 3 min 30 or 40 cycles: –95 °C 1 s –55 °C 30 s	[8, 9, 12, 15, 17, 18]	370 bp fragment with TStaG422/TStag765 412-bp fragment with *tuf* primers

(continued)

Table 1
(continued)

Target	Primers for amplification	PCR amplification conditions	Source/s	Product size
sodA internal gene	*d1* 5′-CCITAYICITAYGAYGCIYTIGARCC-3′ *d2* 5′-ARRTARTAIGCRTGYTCCCAIACRTC-3′ *sodA*-F 5′-CATAACACTTATGT WACTAAATTAAA-3′ *soda*-R 5′-ATCTAAAGAACCCCATTGTTC-3′	95 °C 5 min 30 cycles: –95 °C 1 min –55 °C 1 min –72 °C 1 min 72 °C 10 min	[9, 15, 21]	236 bp fragment based on an internal portion (83 %) of the *sod*A gene
rpoB	Staph rpoB 3554r 5′-CCGTCCCAAGTCATGAACC-3′ Staph rpoB 1418f*5′-CAATTCAT GGACCAAGC-3′	95 °C 5 min 35 cycles: –94 °C 45 s –52 °C 60 s –72 °C 90 s 72 °C 10 min	[9, 19]	899 bp fragment
gap	GF-1 (5′ATGGTTTTGGTAGAA TTGGTCGTTTA-3′) GR-2 (5′-GACATTTCGTTA TCATACCAAGCTG-3′)	94 °C 2 min 40 cycles: –94 °C 20 s –55 °C 30 s –72 °C 40 s 72 °C 5 min	[2, 9, 25, 26]	931 bp fragment

dnaJ / *hsp*40	SA-(F)- 5′-GCCAAAAGAGACTATTATGA-3′ SA-(R) 5′-ATTGYTTACCYG TTTGTGTACC-3′	94 °C 3 min 5 cycles: –94 °C 30 s –45 °C 30 s –72 °C 60 s 30 cycles: –94 °C 30 s –50 °C 30 s –72 °C 60 s 72 °C 3 min	[22]	920 bp fragment
tRNA intergenic spacer	T5A-F 5GTCCGGTGCTCTAACCAACTGAG-3′ T3B-R 5′-AGGTCGCGGGTTCGAATCC-3′	40 cycles: –94 °C 30 s –60 °C 30 s –72 °C 30 s	[16, 24]	50–100 bp

15–20-well 1.5 mm fixed-height gel combs into the comb slots of the gel tray. Make sure that the combs are 1 mm above the bottom of the gel tray so that the wells do not tear. The gel tray should be oriented such that the gel products can migrate toward the red (positive) leads.

5. Add 1 μl ethidium bromide (10 mg/ml; EtBr) to cooling agarose and gently swirl to mix (wear nitrile gloves since EtBr is a carcinogen). Pour the 2 % agarose suspension into the prepared gel tray. Push out bubbles with a disposable tip.
6. Allow the gel to solidify at room temperature for 30 min. After the gel has solidified add 0.5× TBE buffer (running buffer) to the gel rig buffer tank fully submerging the 2 % agarose gel.
7. PCR reaction: 0.2 ml singlet tubes, snap top tubes, strips of tubes, or 96-well format with sealing film.
8. Microcentrifuge tubes (1.5 ml).
9. PCR reagents (*see* Tables 1 and 2).
10. 100 bp DNA ladder.
11. Gel horizontal rig, electrophoresis instrument with power supply.
12. UV transilluminator box or a UV transillumination camera system.
13. Vortex mixer.
14. Loading dye: The final concentration of the loading dye should be 1×; loading dyes can be either made or purchased at 5×, 6×, or 10× concentrations. If working with a 5× loading dye, add 2 μl of the dye to 8 μl of the PCR product. To make a 5× bromophenol blue loading dye weigh out 4 mg of bromophenol blue and mix with 5 ml of 10× TBE buffer and 5 ml glycerol and store in a sealed conical tube at room temperature.
15. Thermocycler.
16. PCR cleanup: For ethanol precipitation of PCR amplification products, prepare 70 or 80 % EtOH in molecular grade H_2O and prepare 3 M sodium acetate by weighing out 204.12 g of sodium acetate trihydrate (NaOAc). Add the 204.12 g to 400 ml of deionized H_2O (dH_2O). Stir until dissolved and then adjust to a pH 4.0 with glacial acetic acid. Adjust the final volume to 500 ml with dH_2O. Isopropanol precipitation requires 100 % isopropanol and a preparation of 7.5 M sodium acetate. Prepare the NaOAc by weighing out 510.3 g of sodium acetate trihydrate (NaOAc). Add the 510.3 g to 400 ml of deionized H_2O (dH_2O). Stir until dissolved and then adjust to a pH 4.0 with glacial acetic acid. Adjust the final volume to 500 ml with dH_2O (*see* **Notes 9** and **10**).

Table 2
PCR reaction buffers

16sRNA [15]	Gap gene [7, 9, 25, 26]	Internal transcribed spacer ITS; 16s–23s spacer regions [7, 13, 20]
PCR buffer (10× supplied with *Taq* polymerase) 10 μl	*DNA* 5 μl	*DNA* 1.25 μl (conc. 20 ng/μl)
dNTPs 200 μM conc. each	*Taq* DNA polymerase 1.25 U	*Gene Amp 10× PCR buffer* 10 μl
Genomic DNA (120 ng/ml) 2 μl	*10× PCR amplification buffer* 5 μl	*dNTP mixture* 200 μM of each
Taq DNA polymerase (0.2 U) 0.2 μl	*Primers* 0.8 μM each	*Primer G1* 0.5 μM
Primers (20 μM) 2.5 μl each	*dNTPs* 0.4 mM each	*Primer L1* 0.5 μM
dH_2O to meet final volume	*dH_2O* to meet final volume	*AmpliTaq DNA polymerase* (Perkin-Elmer) 1.3 U after hot start
Final volume 100 μl	*Final volume* 50 μl	*dH_2O* to meet final volume
		Final volume 100 μl
tDNA-ILP PCR [16, 24]	*tuf* gene [15, 17, 18]	sodA [15]
DNA template 1 ng	*PCR buffer* (10× supplied with *Taq* polymerase) 10 μl	*PCR buffer* (10× supplied with *Taq* polymerase) 10 μl
1× Taq polymerase buffer (50 mM KCl, 10 mM Tris–HCl [pH 8.8 at 25 °C], 1.5 mM $MgCl_2$)	*dNTPs* 200 μM conc. each	*dNTPs* 200 μM conc each
dNTP mixture 0.2 mM each	*Genomic DNA* (120 ng/ml) 2 μl	*Genomic DNA* (120 ng/ml) 2 μl
Primer T5A 0.5 μM	*Taq* DNA polymerase (0.2 U) 0.2 μl	*Taq* DNA polymerase (0.2 U) 0.2 μl
Primer T3B 0.5 μM	Primers (20 μM) 2.5 μl each	*Primers* (20 μM) 2.5 μl each
Taq polymerase 1.25 U	*dH_2O* to meet final volume	*dH_2O* to meet final volume
dH_2O to meet final volume	*Final volume* 100 μl	*Final volume* 100 μl
Final volume 50 μl		

2.2 Real-Time PCR Amplification

1. Instrument examples are LightCycler 480 or LightCycler Red 640 Real-Time PCR System (Roche) or ABI Prism 7300 Instrument (Applied Biosystems).
2. Reagents are listed in Table 3.
3. Microcentrifuge and rotor centrifuge.
4. Blocked/filtered pipette tips.
5. Real-time quality 0.2 ml standard 94-well plates, single tubes, or caps along with appropriate caps or plate seal.

2.3 Fluorescent In Situ Hybridization Reagents

1. 4 % Paraformaldehyde solution: Add 800 ml of 1× PBS to a glass beaker. Place on a magnetic hot stir plate and heat to 60 °C in a ventilated hood. Add 40 g of paraformaldehyde powder while stirring. Add a few drops of NaOH to help the paraformaldehyde dissolve. Once fully dissolved, allow the solution to cool and then add 1× PBS to make 1 l total volume. Check the pH, and add a few drops of HCL if needed for a pH 6.9. Filter-sterilize and store at 4 °C. Use within 1 month. Use caution, and wear gloves and a mask when preparing since paraformaldehyde is toxic.
2. Hybridization solution I: Prepare hybridization solution by combining 0.9 M NaCl, 20 mM Tris–HCl at a pH 7.5, 0.001 % SDS, 43 % formamide, and 50 ng of the Cy3 fluorescein-labeled probe.
3. Humidified chamber: To make a humidity chamber, place wet tissues into the base of a clean plastic pipet-tip box.
4. Hybridization solution II: Combine 10 % (wt/vol) dextran sulfate, 10 mM NaCl, 30 % (vol/vol) formamide, 0.1 % (wt/vol) sodium pyrophosphate, 0.2 % (wt/vol) polyvinylpyrrolidone, 0.2 % (wt/vol) Ficoll, 5 mM disodium EDTA, 0.1 % (vol/vol) Triton X-100, and 50 mM Tris–HCl (pH 7.5).
5. FISH Probe Sep148 sequence 50-AATATATTATCCGGT-30: To prepare the Sep148 PNA Probe (N-terminus is attached to an Alexa fluor 488 dye via a double-AEEA linker [excitation 470–490 nm, emission LP 516 nm]) solubilize the probe in 10 % of acetonitrile and 1 % of trifluoroacetic acid to obtain a 100 μM stock solution (*see* **Note 1**).
6. Washing solution: To prepare the washing solution mix 5 mM Tris base, 15 mM NaCl, and 1 % (vol/vol) Triton X (pH 10).
7. Charged Adhesion Slide (Thermoscience) for both paraffin-embedded tissue sections and for smearing of bacterial colonies.
8. Citifluor solution (Citifluor, Ltd): An anti-bleaching solution.
9. Tryptic soy agar (TSA).
10. Non-fluorescent immersion oil or IMAGEN Mounting Fluid (DAKO).
11. Epifluorescence microscope with one filter sensitive to Alexa Fluor 488.

Table 3
Real-time PCR probe sets

Gene	Primers	Probes	Reaction mixtures	Cycling parameters	Instrument	Source/s
tuf	*STAR1*: 5′-GCGGdTCCA TCTATAAGTGA-3′ *STAF1*: 5′-GGGTGAGT AACACGTGGA-3′	*CNS Probe A*: 5′-GGATAA TATATTGAACCGCA-3′ has a 5 °C annealing temperature above the primers *CNS probe A*: peak 0 has a melting temperature of 58 °C specific to *S. epidermidis*	*STAR1*: 5 pmol/μl *STAF 1*: 1 pmol/μl *DNA*: 10 ng *Tris–HCL (pH 8.3)*: 500 mM $MgCl_2$: 5 mM Bovine serum albumin: 1 mg/ml *dNTPs(Gibco, BRL)*: 200 μM conc. each *Platinum taq polymerase (Gibco BRL)*: 0.4 U *Sybr Green I (Bio/Gene Ltd.)*: diluted 0.01 % *Biprobe*: 5 pmol/μl *Total volume*: 10 μl	94 °C for 5 s 40 cycles (temperature transition rate of 20 °C/s) 94 °C for 0 s 55 °C for 1 s 60 °C for 15 s 74 °C for 10 s Anneal at 55 °C for 1 s, and then take fluorescent reading Follow PCR cycling with a melting curve analysis of 40 °C–95 °C (temperature transition rate of 0.2 °C/s) with continuous fluorescence readings at each transition	LightCycler	[17]
atlE gene *tuf* gene	*altE-F*: 5′-GGAGGAACTA ATAATAAGTTAACTG-3′ *altE-R*: 5′-GTCATAAACA GTTGTATATAAGCC-3′ *Tuf-P1*: 5′-AAACAACTGTT ACTGGTGTAGAAATG-3′ *Tuf-P2*: 5′-AGTACGGA AATAGAATTGTG-3′	*altE Probe*-5′-FAM-CTGCT AATCGTGGTGTTGCT CAAATTAAA-BHQ1-3′ *Tuf Probe*-5′-TEXAS RED-TCCGTAAATT ATTAGACTACGC TGAAGC-BHQ2-3′	*Primers*: 0.8 μM/L each *Fluorophore probe*: 0.4 μM/L $MgCl_2$: 6 mmol/L *dNTPs*: 200 μmol/L *Super Hot Taq Polymerase*: 1I U *Buffer PCR*: 1× *DNA/Lysate*: 1 μl	95 °C for 15 min 35cycles 95 °C for 15 s 60 °C for 60 s Detection, not melting curve based	7500 ABI prism sequence detector (applied biosystems) *altE* specific to *S. epidermidis* *tuf* gene primer set and probe generic for all *Staphylococcus* species is an inhibition control	[27, 28]

(continued)

Table 3
(continued)

Gene	Primers	Probes	Reaction mixtures	Cycling parameters	Instrument	Source/s
16s rRNA	*16S3 up*: 5′-ATGCAAGTCG AGCGAAC-3′ *16S3 down*: 5′-TGTCTCA GTTCCAGTGTGGC-3′ Product length: 282 bp	*16S FITC-probe*: 5′-CACTTA TAGATGGATCCGCGCC GCATTAGCTAGT-FITC-3′ *16S Red-probe*: 5′-Red-GTAAGGTAAC GGCTTACCAAGGC AACGAT-Phos-3′	*Primers*: 0.5 μM final conc *Probes*: 0.2 μM final conc *DNA*: 2 μl *$MgCl_2$*: 3.5 mM final conc *10× LightCycler DNA FastStart master hybridization probe kit mix*: 2 μl *Total volume*: 20 μl	95 °C for 10 min 45 cycles: 95 °C for 10 s 57 °C for 10 s 73 °C for 10 s Melt-curves were generated by holding the capillaries at 95 °C for 10 s, followed by 40 °C for 15 s. The temperature was then increased at 0.2 C/s to 95 °C with step fluorescence acquisition	LightCycler The Tm for *S. epidermidis* was 68.14 °C with SD of 1.08 °C	[14, 23]

2.4 MALDI-TOF-MS Supplies

1. MALDI-TOF-MS instruments common to the clinical microbiology laboratory are the Bruker MicroFlex LT system (Bruker Daltonics) with the MALDI Biotyper automation control and Biotyper Software version 3.0 and the Shimadzu Accuspot/Vitek MS MALDI-TOF MS system and software (bioMérieux); both should be utilized according to the manufacturer's instructions.
2. 24-spot steel plate, 96-spot steel plate, or 384-spot steel target plate.
3. Matrix solution for MALDI-TOF-MS is known as MALDI matrix. It is composed of a saturated solution of α-cyano-4-hydroxycinnamic acid (HCCA) in 50 % acetonitrile and 2.5 % trifluoroacetic acid.
4. Ethanol 70 %.
5. Bacterial test standard for MALDI-TOF-MS is needed for calibration and validation of the instrument. There are two options; either the Bacterial test standard (Bruker Daltonics) or the external calibration panel of angiotension, glu-fibrinopeptide B, adrenocorticotropic hormone, bovine insulin, ubiquitin, and cytochrome C can be utilized as a 1:1 (v/v) mixture of matrix solution and calibration peptide (0.25 pmol per spot).
6. Agar for culturing bacteria: Mueller-Hinton agar or TSA.

3 Methods

3.1 PCR Amplification and Sequencing

16s rRNA gene amplification with sequencing is commonly utilized to identify bacteria; however, the 16s rRNA gene is not as discriminatory within the CONS species [7–9, 11–26]. Table 1 contains various primers for genes that have successfully been shown to differentiate *S. epidermidis* from other CONS. Refer to Table 1 for primer sequences, PCR amplification parameters, and approximate PCR product size [2, 7–9, 11, 13–15, 19–22]. Perform all PCR reaction preparations on ice. Change pipette tips between PCR products, reagents, and genomic DNA samples.

1. After selecting the appropriate primer set from Table 1, prepare the PCR reaction mixture (cocktail) from Table 2 [7, 9, 13, 15–18, 20, 24–26]. To do this, determine the number of samples that will be amplified including a negative control, positive control, and an extra reaction to account for pipetting error. Next, multiply the volume of each reagent by the number of reactions. Combine all reagents, except for the DNA, at the correct volume in a 1.5 ml microcentrifuge tube (sterile), adding the polymerase last. Mix by flicking the side of the tube. Briefly centrifuge for 15–30 s in a microcentrifuge at a low rpm to collect all reagents together in the bottom of the tube.

2. Add the indicated volume of the cocktail to each PCR reaction tube (0.2 ml reaction tubes). For example, if the reaction volume is 50 μl total then add 45 μl of the cocktail and 5 μl of the genomic DNA or bacterial lysate. PCR reaction tubes can either be single tubes, strips of tubes, or a 96-well plate format.
3. Add the genomic DNA or lysate to the designated PCR reaction tube. Close the lids or seal the top of the 96-well plate with the appropriate sealed film. Briefly (15–30 s) centrifuge the reaction tubes.
4. Load samples onto the thermocycler, making sure that the tube caps are securely sealed.
5. Program the thermocycler with the appropriate cycling parameters listed in Table 1.
6. Store the amplified product at 4 °C (short term <24 h or at −80 °C long term).
7. To verify that amplification occurred make a 2 % agarose gel as described in Subheading 2.4 (*see* **Note 2**).
8. Using 1 or 1.5 ml microcentrifuge tubes, add PCR product and the loading dye to a final volume of 10–12 μl (1 volume of loading dye to 5 volume of purified DNA). Properly label the tubes so that samples do not get swapped.
9. Briefly centrifuge the microcentrifuge tubes for 15–30 s.
10. Add 3–5 μl of a 100 bp DNA ladder to a single well of the 2 % agarose gel. There should be at least one well containing DNA ladder per row of wells in the agarose gel. The ladder confirms that the gel ran properly and allows measurement of the PCR product band sizes.
11. Inoculate one PCR product per well on the agarose gel. Record loading order of the products.
12. The gel needs to be oriented in the gel rig such that the products will run toward the positive red lead (RUN TO RED). Place the safety cover over the gel rig. Attach the positive (RED) and negative (BLACK) leads to the power supply. Turn on the power supply.
13. Set the voltage and time on the power supply. Then press the run button. If the fragment is 50–150 bp in length, run for 30 min at 90–100 V when using a 2 % gel. Large fragments will run slower than smaller fragments. Adjust run-time accordingly. Beware of running the gel too long or the smaller fragments will migrate out of the gel.
14. To check for the presence of PCR product bands use a ultraviolet light source or a UV-camera box. If using a UV box, wear a proper face-shield/eye protection since UV light will damage eyes. The EtBr will intercalate into the DNA and fluo-

resce in the presence of UV light. Wearing gloves, remove the gel tray from the gel rig and place on the UV light source. Turn on UV light source. The bands should appear orange in the gel on a UV light box, or if using a camera box with UV transillumination they will appear as white bands in a black gel or as black bands in a grayish gel, depending on the camera parameters. Be sure to properly dispose of the agarose gel and the running buffer since EtBr was utilized.

15. Ethanol (a) or isopropanol (b) precipitation can be utilized to purify PCR products prior to sequencing.
    (a) Ethanol precipitation: In a properly labeled 1.5 ml microcentrifuge tube, add 3 volumes of 70–80 % ethanol (EtOH) and 1/10 volume of 3 M sodium acetate (NaOAc) to 1 volume of the PCR product. Briefly vortex for 15–30 s. Place suspension in a −70 °C freezer for 20 min. Centrifuge sample in a microcentrifuge for 10 min to pellet the DNA. A pellet should be visible in the bottom of the tube (if not visible re-centrifuge). Next, perform the wash step by adding 1 volume of 70–80 % EtOH. Vortex the sample for 15–30 s. Centrifuge for 1 min. Again, visually verify the presence of a pellet. If no pellet is present, re-centrifuge for an additional 10 min. Aspirate off the EtOH. Allow the samples to dry in ambient air at room temperature by inverting the microcentrifuge tubes over clean absorbent paper. Resuspend the DNA pellet in molecular grade water to desired volume (30–50 μl). Store purified DNA at 4 °C or −80 °C until ready to sequence.
    (b) Isopropanol precipitation: In a 1.5 ml microcentrifuge tube, add 2 volumes of isopropanol and 1/2 volume of 7.5 M sodium acetate (NaOAc) to 1 volume of PCR product. Briefly vortex for 15–30 s. Incubate at room temperature for 10 min. Centrifuge sample in a microcentrifuge for 10 min to pellet the DNA. A pellet should be visible in the bottom of the tube (if not visible re-centrifuge). Perform the wash step by adding 1 volume of 80 % EtOH. Vortex the sample for 15–30 s. Centrifuge for 1 min. Visually verify the presence of a pellet. If no pellet is present, re-centrifuge for an additional 10 min. Aspirate off EtOH. Allow the samples to dry in ambient air at room temperature by inverting the microcentrifuge tubes over absorbent paper. Then resuspend the DNA pellet in molecular grade water to desired volume (30–50 μl). Store purified DNA at 4 °C or −80 °C until ready to sequence.
16. Sequencing is dependent upon the sequencer platform available. This discussion will not describe the specific methodology [2, 7, 8, 12–26]. *See* Table 4 for the sequencing primers that correspond to those of the amplified product.

Table 4
Sequencing primers

Sequencing targets	Primers for sequencing	Source/s
16 s rRNA gene	SSU-bact-27f 5′-AGAGTTTGATCMTGGCTCAG-3′ SSU-bact-519R 5′-GWATTACCGCGGCKGCTG-3′ Epsilon F 5′-AAGAGTTTGATCCTGGCTCAG-3′ 320R 5′-TTGACCGTGTCTCAGTTCCA-3′	[9, 14, 15, 23]
16S–23S internal transcribed spacer sequence	G1(F):: 5′-GAAGTCGTAACAAGG-3′ L1(R) 5′-CAAGGCATCCACCGT-3′	[7, 13, 20]
tuf	TStaG422 (F) 5′-GGCCGTGTTGAACGTGGTCAAATCA-3′ TStag765 (R) 5′-TIACCATTTCAGTACCTTCTGGTAA-3′ *tuf*-F 5′-GCCAGTTGAGGACGTATTCT-3′ *tuf*-R 5′-CCATTTCAGTACCTTCTGGTAA-3′	[8, 9, 12, 15, 17, 18]
sodA internal gene	*d1* 5′-CCITAYICITAYGAYGCIYTIGARCC-3′ *d2* 5′-ARRTARTAIGCRTGYTCCCAIACRTC-3′ *sodA*-F 5′-CATAACACTTATGTWACTAAATTAAA-3′ *soda*-R 5′-ATCTAAAGAACCCCATTGTTC-3′	[9, 15, 21]
rpoB	Staph rpoB 1418f*5′-CAATTCATGGACCAAGC-3′ Staph rpoB 1975r* GCIACITGITCCATACCTGT-3′ Staph rpoB 1876r* 5′-GAGTCATCITTYTCTAAGAATGG-3′	[9, 19]
gap	Gap1-for 5′-ATGGTTTTGGTAGAATTGGTCGTTTA-3′ Gap3-rev 5′-G(ACT)TTT(AGCT)A(CT)TTCTT(AGT)-3′ Gap4-for 5′-GA(CT)GT(AGCT)GT(AGCT)(CT)T(AT) GAATGTAC(AT)GG-3′ Gap5-rev 5′-GTT(AT)GT(AT)GTACA(AGT)GA(ACT) GCACC(AT)G-3′ Gap6-for 5′-AAGG(ACT)(CT)T(ACT)ATGAC(AGT) AC(AT)AT(CT)CA(CT)G-3′ Gap7-rev 5′-GAACC(AT)GT(AT)GC(AT)AC(AT) GG(ACT)ACACGTTG-3′ Gap8-for 5′-GAA(ACT)CATT(CT)GGTTACA(AC)(ACT) GA(AT)GA(CT)G-3′ Gap2-rev 5′-GACATTTCGTTATCATACCAAGCTG-3′	[2, 9, 25, 26]
dnaJ/*hsp*40	SA-(F)- 5′-GCCAAAAGAGACTATTATGA-3′ SA-(R) 5′-ATTGYTTACCYGTTTGTGTACC-3′	[22]
tRNA intergenic spacer	T5A-F 5′-GTCCGGTGCTCTAACCAACTGAG-3′ The 5′ end was labeled with Cy5	[16, 24]

17. Compile completed sequences into contigs using software such as ChromasPro (Technelyslum), DNA STAR (DNA STAR, Inc.), or Clustal Omega (EMBL-EBI).
18. Compare completed sequences to those of a credible database such as published in Genbank by Blast, EMBL Data Library Nucleotide sequence database collaboration, DDBJ DNA Database of Japan, or RIDOM (version 1.2, Ridom GmbH).

19. When querying sequence results to those of the database, the query match percentage needs to be >98 %. A high match is needed to identify the isolate to the species level.

3.2 Real-Time PCR

Another technique to identify *Staphylococcus epidermidis* is real-time PCR alone or in conjunction with melt-curve analysis. Three optional gene targets are listed in Table 3 [14, 17, 23, 27, 28] along with an extraction and amplification control gene *tuf*. Assemble all real-time PCR reaction preparations on ice. Use sterile blocked pipette tips (*see* **Note 12**).

1. Thaw all components at room temperature. Mix gently and then centrifuge at 4 °C to collect contents in the bottom of the tube. Prepare the PCR primers and probes with molecular grade H_2O. Place on ice.
2. Calculate the number of reactions needed for each primer set (including two positive, four negative no-template controls, and a nonspecific control which will consist of target DNA of a known *Staphylococcus aureus* strain). The four non-template controls will consist of molecular grade H_2O for two controls and reaction buffer in two wells instead of DNA template. In a 1.5 ml microcentrifuge tube (on ice), prepare the real-time PCR reaction cocktail based on Table 3 for the total number of reactions.
3. Gently mix the reaction cocktail by pipetting up and down. Do not vortex. Dispense equal aliquots of the recommended volume of the reaction cocktail to each reaction well or tube. Use caution to avoid pipetting errors.
4. Add the recommended volume of DNA to the reaction tubes. Add molecular grade water to two of the negative control wells in place of target DNA, and then add reaction buffer to the two other negative control wells in place of target DNA. Do not cross contaminate the samples.
5. *S. epidermidis*-positive control strains: ATCC 35984 and ATCC 14990 strains. Control strains should be extracted like that of test isolates. Add the positive control DNA template to the positive control wells or tubes.
6. Add the target DNA either extracted or lysate to each experimental reaction tube. Ideally, a 5–10 μl of target DNA improves assay precision. Perform triplicate repeats for each isolate.
7. If using a plate format, cover the reaction plate with adhesive cover. If using thin-wall tubes, properly close/seal the tops of the tubes.
8. Centrifuge the reaction plate or thin-wall tubes briefly to collect all of the liquid in the bottom of the wells/tubes at 1,500 × *g* for 5 s (3,000 rpm in a standard swing-bucket

centrifuge containing a rotor for 96-well plates with suitable adaptors). Balance the centrifuge. Spin the reaction tubes to transfer the master mix to the bottom of the tube, and insure that no bubbles form in the reaction mixture.

9. Turn on the real-time instrument (see operations manual). Load in the plate/tubes. Assign the location of samples on the instrument. Start the real-time PCR program, and perform a melting curve analysis. Remember that the program cycling parameters are available in Table 3 (*see* **Notes 3–6**). The melting curve analysis is performed after the last amplification cycle. Select the appropriate fluorescent detectors (*see* Table 3) and tasks per well/tube. Make sure to assign wells as negative control, positive control, nonspecific control, or unknown (isolate number).
10. A positive is determined when a cycle threshold (Ct) value is above the threshold. (*see* **Note 13**) If a melt curve analysis is being performed, then a positive unknown isolate that produces a characteristic melting curve with a discernible peak at the Tm (melting temperature) that is comparable to the positive control Tm is identified as the same species as the positive control [29] (*see* **Note 3**). The Ct is the cycle of the PCR protocol in which significant fluorescence appears. There should be 0 Ct for the negative control well. If the negative control fails then the assay must be repeated. The positive control well should have the optimal Ct score and melting temperature (Tm) for the probe. The *tuf* gene can be used as an extraction and amplification control in the same tube if using different fluorescent markers. A read should be taken prior to PCR amplification, during the PCR cycling, and also at the endpoint. The qualitative real-time PCR interpretations will be yes or no. Confirm the identity of the isolate based on the specific peak Ct and Tm of the unknown (average of isolate ran in triplicate) to the Ct and Tm of the positive control (average of control ran in triplicate).

3.3 Fluorescent In Situ Hybridization

1. FISH of paraffin-embedded tissue samples: Section paraffin-embedded tissue, with detectable gram-positive cocci in clusters by Gram-stain, to 10–20 μm and mount onto a charged adhesion slide. Also include control slides of *S. aureus* (negative) and *S. epidermidis* (positive). Remove the paraffin by treating sections with Rotihistol (100 %) (Roth) for 10 min. Follow with two 10-min washes with a solution of Rotihistol–ethanol at a 1:1 dilution. Add 15 μl of hybridization solution I (*see* Subheading 2.3, **item 2**) and 500 nM Sep148 PNA probe specific to the tissue on the microscope slide. Incubate for 3 h at 43 °C in a humidified chamber. Wash at 43 °C with distilled H_2O to remove unbound probe, air-dry, and mount with Citifluor solution (store in a dark drawer). Fluorescence is detected by using an epifluores-

cence microscope (*see* Subheading 2.3, **item 11**) equipped with one filter sensitive to the Alexa Fluor 488 signaling molecule attached to the PNA probe (*see* **Note 7**).

2. To identify *S. epidermidis* from isolated colonies smeared onto a microscopic slide: Inoculate TSA with the bacterial isolate and incubate at 37 °C for 24 h. Select colonies from the plate and suspend in a small volume (50 μl) of sterile molecular grade H_2O. Homogenize suspension by vortexing for 1 min. Smear a few drops onto a glass microscope slide. Fix suspension onto the microscope slide by immersing in 4 % (wt/vol) paraformaldehyde (*see* Subheading 2.3, **item 2**) (use caution, and wear gloves and a mask when preparing since paraformaldehyde is toxic) and then transferring to a container of 50 % (vol/vol) of EtOH for 10 min. Allow smeared slides to air-dry. Then cover with 20 μl of the hybridization solution and 500 nM of the Sep148 PNA probe [30]. Add hybridization solution II (*see* Subheading 2.3, **item 4**) [31]. Cover samples with cover slips, place in a moist humidity chamber, and incubate for 90 min at 50 °C. Remove the cover slip, and submerge the slide in washing solution (Subheading 2.3, **item 6**) that has been pre-warmed to 50 °C. Wash for 30 min at 50 °C. Allow slides to air-dry. Add one drop of non-fluorescent immersion oil (*see* Subheading 2.3, **item 10**), and cover slip the slide. View slides within 24 h (store in a dark drawer). Fluorescence is detected by using an epifluorescence microscope (examples are listed in Subheading 2.3, **item 11**).

3.4 MALDI-TOF-MS

MALDI-TOF-MS [8, 32–34]: MALDI-TOF-MS is being implemented within the clinical microbiology laboratory for rapid identification to the species level based upon the unique peptide peak fingerprint. Two commercial instruments that are commonly used for this identification are listed in Subheading 2.4, **item 1**; these are utilized in conjunction with peptide databases that contain extensive peptide fingerprint libraries for clinically relevant bacteria (*see* **Note 11**).

1. Inoculate isolates of interest onto Mueller-Hinton agar, Columbia agar supplemented with 5 % horse blood, or nutrient agar.
2. Incubate for 24–48 h at 37 °C.
3. Prepare the target plates by wiping with methanol and allow to air-dry.
4. There are three optional methods for plate spotting:
5. Select an isolated colony with a calibrated 1 μl calibrated loop, emulsify in 100 μl of sterile molecular grade H_2O, and spot 1 μl of the emulsion in triplicate onto a steel target plate (*see* **Note 8**).

6. For the direct transfer procedure (DTP), inoculate the colony with a 1 μl calibrated loop directly onto three wells (in triplicate), on the steel target plate.
7. To perform the formic acid extraction procedure, select a large isolated colony and resuspend in 300 μl HPLC-grade dH_2O in a microcentrifuge tube. Add 900 μl EtOH. Centrifuge for 2 min, then remove the supernatant, and air-dry the pellet. Resuspend in 50 μl of 70 % formic acid and 50 μl of 100 % acetonitrile. Centrifuge for 2 min. Add 1 μl of the aliquot of supernatant to three wells of the steel target plate.
8. Dry the spotted target plate for 5 min at room temperature. Add 1 μl of absolute ethanol to each well of the plate.
9. Allow target wells to dry, and then overlay each well with 1 μl of matrix solution (*see* Subheading 2.4, **item 3**). Dry in a fume hood.
10. Calibrate the instrument with the protein calibration standard described in Subheading 2.4, **item 5**.
11. Then load target plate into the MALDI-TOF-MS of choice as per the manufacturer's instructions. Using the instrument software, specify the target positions for the analyte in each well. Perform the analysis by following the manufacturer's settings. Example settings are an accelerating voltage of 20 Hz in linear mode with each spectrum the sum of the ions obtained from 200 laser shots performed in five different regions of the same well. The spectra are analyzed at an m/z range of 1,000–11,000 [8, 33, 34]. Next, compile the data of three replicates to minimize random variation, and then compare the fingerprint of the isolate to that of the database, which corresponds to the instrument of choice.
12. For confirmation of the identity a score of ≥2.0 is specific to the species level for the procedure described in 5A and 5B. For the procedure described in 5C a score of ≥1.7 is specific to the species level [35].

4 Notes

1. Regarding PNA FISH, a solubilization solution (10 % acetonitrile and 1 % trifluoroacetic acid) was essential to the functionality of the probe; this solution worked, whereas H_2O, dimethylformamide, acetonitrile, or trifluoroacetic acid alone did not sufficiently solubilize the probe.
2. To minimize preparation and disposal, pre-made agarose gels such as the E-Gel can be purchased from a supplier such as Life Technologies Corporation.

3. Regarding PCR amplification, the reaction annealing conditions should be adjusted based on the primer Tm (melting temperature). The exact Tm will be supplied by the manufacturer. This is particularly important when double bands are detected instead of a single product.
4. Regarding PCR amplification, the elongation time should be adjusted based on 60 s per 1 Kb. A stringent elongation step can help to minimize nonspecific binding.
5. Addition of 2–6 % dimethyl sulfoxide (DMSO) to the PCR reaction can help to reduce nonspecific binding. The annealing temperature should be decreased by 1 or 2 °C when DMSO is utilized. Wear nitrile gloves since DMSO will penetrate latex gloves and can be absorbed into skin.
6. All equipment should be cleaned with a product that denatures RNases and DNases.
7. Regarding FISH, when examining samples utilized the remaining filters on the epifluorescent microscope to check for autofluorescence of the probe stained cells. If significant fluorescence is detected under the other filters then it indicates that the slides are compromised and need to be repeated. Additionally, in these situations it may be helpful to filter sterilize all solution. Do not let the dry slides dry out too long or the chance for autofluorescence increases substantially.
8. MALDI-TOF-MS for species level identification. If only one well is spotted per isolate, the top score must be 10% greater than all other matches in order to speciate the isolate.
9. When performing PCR cleanup by the EtOH precipitation method, the concentration of DNA can be increased by extending the freezer time by 30–60 min.
10. Commercial assays for PCR cleanup have been described, but a few examples are the DNA clean and concentrator 25 kit by Zymo Research (purifies up to 25 μg of DNA; the Qiaquick PCR purification kit (recovery range 100 bp–10 kb) and Agencourt AMPure XP—PCR Purification (recovery range 100 bp and greater).
11. In cases where a proprietary database is not available for MALDI-TOF-MS analysis, the peptide fingerprint profile can be analyzed and compared to the positive control wells by using the software BGP/database available at http://bgp.sourceforge.net/.
12. Sybr green can be used to monitor the development of nonspecific ds-DNA and also to determine the melting curve of a target. It is light sensitive, so keep out of direct light.
13. Since this real-time assay is a qualitative and not a quantitative real-time PCR assay, the DNA standard curve is optional. To create a DNA standard curve for the Ct, use a control strain of *S. epidermidis* at a 1×10^8 CFU. Extract the DNA per a

standard protocol. Perform ¼ dilutions by diluting the standard with H_2O, so the curve will be 1:4, 1:16, 1:24, 1:64, and 1:256 wells. Run these with target primers and probes using the same reaction mixture as the experimental isolates.

References

1. Zadoks RN, Watts JL (2009) Species identification of coagulase-negative staphylococci: genotyping is superior to phenotyping. Vet Microbiol 134:20–28
2. Sheraba NS, Yassin AS, Amin MA (2010) High-throughput molecular identification of Staphylococcus spp. isolated from a clean room facility in an environmental monitoring program. BMC Res Notes 3:278
3. Patteet L, Goossens H, Ieven M (2012) Validation of the MicroScan-96 for the species identification and methicillin susceptibility testing of clinical significant coagulase-negative staphylococci. Eur J Clin Microbiol Infect Dis 31:747–751
4. Renneberg J, Rieneck K, Gutschik E (1995) Evaluation of Staph ID 32 system and Staph-Zym system for identification of coagulase-negative staphylococci. J Clin Microbiol 33:1150–1153
5. Becker K, Von Eiff C (2011) *Staphylococcus*, *Micrococcus*, and other catalase-positive cocci. In: Versalovic J, Carroll KC, Funke G, Jorgensen JH, Landry ML, Warnock DW (eds) Manual of Clinical Microbiology, 10th ed. ASM Press, Washington, DC, pp 308–330
6. Kloos WE, Wolfshohl JF (1982) Identification of *Staphylococcus* species with the API STAPH-IDENT system. J Clin Microbiol 16: 509–516
7. Jensen MA, Webster JA, Straus N (1993) Rapid identification of bacteria on the basis of polymerase chain reaction-amplified ribosomal DNA spacer polymorphisms. Appl Environ Microbiol 59:945–952
8. Carpaij N, Willems RJ, Bonten MJ et al (2011) Comparison of the identification of coagulase-negative staphylococci by matrix-assisted laser desorption ionization time-of-flight mass spectrometry and *tuf* sequencing. Eur J Clin Microbiol Infect Dis 30:1169–1172
9. Ghebremedhin B, Layer F, Konig W et al (2008) Genetic classification and distinguishing of *Staphylococcus* species based on different partial gap, 16S rRNA, *hsp60*, *rpoB*, *sodA*, and *tuf* gene sequences. J Clin Microbiol 46:1019–1025
10. Kwok AY, Chow AW (2003) Phylogenetic study of *Staphylococcus* and *Macrococcus* species based on partial *hsp60* gene sequences. Int J Syst Evol Microbiol 53:87–92
11. Becker K, Harmsen D, Mellmann A et al (2004) Development and evaluation of a quality-controlled ribosomal sequence database for 16S ribosomal DNA-based identification of *Staphylococcus* species. J Clin Microbiol 42:4988–4995
12. Bergeron M, Dauwalder O, Gouy M et al (2011) Species identification of staphylococci by amplification and sequencing of the tuf gene compared to the gap gene and by matrix-assisted laser desorption ionization time-of-flight mass spectrometry. Eur J Clin Microbiol Infect Dis 30:343–354
13. Couto I, Pereira S, Miragaia M et al (2001) Identification of clinical staphylococcal isolates from humans by internal transcribed spacer PCR. J Clin Microbiol 39:3099–3103
14. Edwards KJ, Kaufmann ME, Saunders NA (2001) Rapid and accurate identification of coagulase-negative staphylococci by real-time PCR. J Clin Microbiol 39:3047–3051
15. Heikens E, Fleer A, Paauw A et al (2005) Comparison of genotypic and phenotypic methods for species-level identification of clinical isolates of coagulase-negative staphylococci. J Clin Microbiol 43:2286–2290
16. Maes N, De Gheldre Y, De Ryck R et al (1997) Rapid and accurate identification of *Staphylococcus* species by tRNA intergenic spacer length polymorphism analysis. J Clin Microbiol 35:2477–2481
17. Martineau F, Picard FJ, Ke D et al (2001) Development of a PCR assay for identification of staphylococci at genus and species levels. J Clin Microbiol 39:2541–2547
18. Martineau F, Picard FJ, Roy PH et al (1998) Species-specific and ubiquitous-DNA-based assays for rapid identification of *Staphylococcus aureus*. J Clin Microbiol 36:618–623
19. Mellmann A, Becker K, von Eiff C et al (2006) Sequencing and staphylococci identification. Emerg Infect Dis 12:333–336
20. Mendoza M, Meugnier H, Bes M et al (1998) Identification of *Staphylococcus* species by 16S-23S rDNA intergenic spacer PCR analysis. Int J Syst Bacteriol 48:1049–1055
21. Poyart C, Quesne G, Boumaila C et al (2001) Rapid and accurate species-level identification of coagulase-negative staphylococci by using the *sodA* gene as a target. J Clin Microbiol 39:4296–4301

22. Shah MM, Iihara H, Noda M et al (2007) *dnaJ* gene sequence-based assay for species identification and phylogenetic grouping in the genus *Staphylococcus*. Int J Syst Evol Microbiol 57:25–30
23. Skow A, Mangold KA, Tajuddin M et al (2005) Species-level identification of staphylococcal isolates by real-time PCR and melt curve analysis. J Clin Microbiol 43: 2876–2880
24. Welsh J, McClelland M (1992) PCR-amplified length polymorphisms in tRNA intergenic spacers for categorizing staphylococci. Mol Microbiol 6:1673–1680
25. Yugueros J, Temprano A, Berzal B et al (2000) Glyceraldehyde-3-phosphate dehydrogenase-encoding gene as a useful taxonomic tool for *Staphylococcus* spp. J Clin Microbiol 38: 4351–4355
26. Yugueros J, Temprano A, Sanchez M et al (2001) Identification of *Staphylococcus* spp. by PCR-restriction fragment length polymorphism of gap gene. J Clin Microbiol 39:3693–3695
27. Kilic A, Basustaoglu AC (2011) Double triplex real-time PCR assay for simultaneous detection of *Staphylococcus aureus*, *Staphylococcus epidermidis*, *Staphylococcus hominis*, and *Staphylococcus haemolyticus* and determination of their methicillin resistance directly from positive blood culture bottles. Res Microbiol 162:1060–1066
28. Kilic A, Muldrew KL, Tang YW et al (2010) Triplex real-time polymerase chain reaction assay for simultaneous detection of *Staphylococcus aureus* and coagulase-negative staphylococci and determination of methicillin resistance directly from positive blood culture bottles. Diagn Microbiol Infect Dis 66: 349–355
29. Kosters K, Reischl U, Schmetz J et al (2002) Real-time LightCycler PCR for detection and discrimination of *Bordetella pertussis* and *Bordetella parapertussis*. J Clin Microbiol 40:1719–1722
30. Stender H, Lund K, Petersen KH et al (1999) Fluorescence In situ hybridization assay using peptide nucleic acid probes for differentiation between tuberculous and nontuberculous *Mycobacterium* species in smears of *Mycobacterium* cultures. J Clin Microbiol 37: 2760–2765
31. Zakrzewska-Czerwinska J, Gaszewska-Mastalarz A, Pulverer G et al (1992) Identification of *Staphylococcus epidermidis* using a 16S rRNA-directed oligonucleotide probe. FEMS Microbiol Lett 79:51–58
32. Carbonnelle E, Beretti JL, Cottyn S et al (2007) Rapid identification of Staphylococci isolated in clinical microbiology laboratories by matrix-assisted laser desorption ionization-time of flight mass spectrometry. J Clin Microbiol 45:2156–2161
33. Rajakaruna L, Hallas G, Molenaar L et al (2009) High throughput identification of clinical isolates of *Staphylococcus aureus* using MALDI-TOF-MS of intact cells. Infect Genet Evol 9:507–513
34. Seng P, Drancourt M, Gouriet F et al (2009) Ongoing revolution in bacteriology: routine identification of bacteria by matrix-assisted laser desorption ionization time-of-flight mass spectrometry. Clin Infect Dis 49:543–551
35. Cherkaoui A, Hibbs J, Emonet S et al (2010) Comparison of two matrix-assisted laser desorption ionization-time of flight mass spectrometry methods with conventional phenotypic identification for routine identification of bacteria to the species level. J Clin Microbiol 48:1169–1175

Chapter 4

Pulsed Field Gel Electrophoresis of *Staphylococcus epidermidis*

Richard V. Goering and Paul D. Fey

Abstract

Pulsed field gel electrophoresis (PFGE) is one of the older methods for the molecular characterization and comparison of microorganisms including bacteria. Nevertheless, PFGE continues to be recognized as the gold standard for molecular typing due to output spanning >90 % of the bacterial genome and standardized protocols and reagents applicable to a wide range of organisms including *S. epidermidis*.

Key words *Staphylococcus epidermidis*, Epidemiology, Molecular typing, Clinical, PFGE

1 Introduction

A variety of molecular methods have been described for the epidemiological analysis of problem pathogens such as *Staphylococcus epidermidis*. However, regardless of their approach, the common goal is to assess epidemiological relatedness based on chromosomal comparisons [1]. Advances in whole-genome sequencing (WGS) are clearly on the horizon as the most direct means of achieving such assessment. However, it is important to note that WGS is currently challenged by the accuracy of relatively short read lengths especially involving repeat sequences [e.g., clustered regularly interspaced short palindromic repeats (CRISPRs), homopolymers, and variable-number tandem repeats (VNTRs)] [2]. An additional issue is the bioinformatics required for the proper annotation and analysis of WGS data which may cost more, in time and money, than the actual sequencing [3, 4]. As WGS continues to be optimized, pulsed field gel electrophoresis (PFGE) remains the overall gold standard for assessing epidemiological interrelationships of pathogens including *S. epidermidis*. The remarkable longevity of PFGE usefulness is due in part to its ability to provide visual genomic comparisons representing >90 % of the bacterial chromosome [1]. In addition, refinements in the preparation of DNA for PFGE analysis have shortened what was initially a multi-day process

Paul D. Fey (ed.), *Staphylococcus Epidermidis: Methods and Protocols*, Methods in Molecular Biology, vol. 1106,
DOI 10.1007/978-1-62703-736-5_4, © Springer Science+Business Media, LLC 2014

to a procedure that can be accomplished in one day. The goal of this chapter is to provide a stepwise description of an optimized protocol for the preparation of *S. epidermidis* chromosomal DNA for PFGE analysis. Appropriate PFGE switching conditions are also detailed as well as comments regarding the visualization, analysis, and interpretation of PFGE data.

2 Materials

2.1 Genomic DNA Extraction

1. Brain heart infusion (BHI) or tryptic soy agar (TSA) plates.
2. Sterile type 1 water.
3. Spectrophotometer.
4. Stationary and shaker water baths (55 °C).
5. BHI or TSA plates.
6. TEN buffer (pH 7.5): 0.1 M Tris–HCl (pH 7.5), 0.15 M NaCl, 0.1 M EDTA.
7. Lysostaphin: 0.5-mg/mL stock in 20 mM sodium acetate (pH 4.5).
8. EC buffer (pH 7.5): 6 mM Tris hydrochloride, 1 M NaCl, 100 mM EDTA, 0.5 % Brij 58, 0.2 % deoxycholate, 0.5 % sodium lauroyl sarcosine.
9. SeaKem Gold agarose (Lonza).
10. TE buffer (pH 8.0): 10 mM Tris-HCl, 1 mM EDTA.
11. PFGE plug mold (BioRad).
12. 15 mL capped culture tubes.

2.2 Restriction Enzyme Digestion

1. 0.5 mL microcentrifuge tubes.
2. Sterile reagent-grade water.
3. SmaI restriction enzyme and 10× buffer (NEB, Roche, etc.)

2.3 PFGE Separation of Chromosomal Macro-Restriction Fragments

1. SeaKem Gold agarose (Lonza).
2. Sterile type 1 water.
3. 0.5× Tris–borate–EDTA (TBE) buffer (pH 8.3): 0.45 M Tris base, 0.45 M boric acid, 15 mM EDTA.
4. BioRad CHEF DRII, DRIII, or Chef Mapper PFGE system.
5. PFGE gel casting frame and comb (BioRad).

2.4 Data Visualization, Capture, and Analysis

1 10 mg/mL ethidium bromide (*see* **Note 1**).

2 Reagent-grade water.

3 Staining box ca. 14 cm × 24 cm.

4 Gel Doc documentation system (BioRad) or equivalent.
5 GelCompar or BioNumerics analysis software (Applied Maths) or equivalent (*see* **Note 2**).

3 Methods

3.1 Genomic DNA Extraction

1. Prepare 2.0 % SeaKem Gold agarose by adding 0.1 g agarose to 5 mL EC buffer and boil until dissolved. Hold at 55 °C until ready to use.
2. Bacterial isolates are grown overnight at 37 °C on BHI or TSA plates.
3. Adjust bacterial cell suspensions in TEN buffer to an absorbance (Optical Density) at 540 nm of 1.0.
4. Harvest 1 mL of cell suspension in a 1.5 mL microcentrifuge at 14,000 × *g* for 20 s.
5. Resuspend the cell pellet in 0.2 mL EC buffer and hold at 55 °C until ready to use.
6. Add 0.2 mL of agarose/EC to each tube, immediately followed by 60 μL lysostaphin solution. Quickly mix by vortexing and cast into PFGE plug molds (two plugs per culture).
7. Allow the plugs to solidify at room temperature for ca. 10 min followed by transfer to 15 mL capped culture tubes containing 5 mL of EC buffer and incubation for 1 h at 37 °C.
8. Replace the EC buffer with 5 mL of pre-warmed TE buffer and incubate at 55 °C with shaking for 1 h with at least two exchanges. After a final exchange of TE buffer, the plugs may be stored at 4 °C until used for restriction enzyme digestion.

3.2 Restriction Endonuclease Digestion of DNA in Agarose Plugs

1. For each bacterial culture, prepare a 1.5 mL microcentrifuge tube to contain 200 μL of restriction enzyme buffer diluted to 1× with sterile reagent-grade water and 4–5 μL of SmaI restriction enzyme (10 U/μL).
2. Transfer a 1.0–2.0 mm thick slice of an agarose plug to each tube and incubate at room temperature for at least 2 h. DNA size standards requiring restriction enzyme digestion should also be prepared at this time (*see* **Note 3**).

3.3 Agarose Gel Casting and PFGE

1. Assemble the PFGE casting frame. Prepare 1.0 % SeaKem Gold agarose by adding 1.0 g to 100 mL of 0.5× TBE and boiling until dissolved. Hold at 55 °C until ready to use.
2. Transfer the DNA-containing agarose gel slices to the PFGE casting comb. If used, DNA size standards should be placed on at least the first and last comb positions. Allow to briefly dry; excess buffer can be blotted from the agarose gel slices using

Kimwipes or other scientific cleaning wipes. Insert the comb into the casting frame, and gently add the molten agarose to completely cover the frame and agarose plugs. Allow to solidify for ca. 30 min at room temperature and then gently remove the comb (*see* **Note 4**).

3. Transfer the agarose gel to the PFGE electrophoresis chamber. Add 2 L 0.5× TBE and prepare for PFGE at 6 V/cm, 120° reorientation angle, 14 °C with switching for a total of 23 h in 2 blocks from 5 to 15 s for 10 h followed by 15–60 s for 13 h. [5] (*see* **Note 5**).

3.4 Gel Staining and Documentation

1. Remove the gel from the chamber and stain with ethidium bromide (1.0 μg/mL final concentration in distilled water) for 20–30 min in a covered container. Destain the gel in distilled water for 30–45 min, and dispose of the ethidium bromide according to institutional procedures for hazardous waste (*see* **Note 6**).
2. Capture the destained gel image using a Gel Doc documentation system (BioRad) or equivalent. PFGE pattern analysis is discussed below (*see* **Note 7**).

4 Notes

1. Staining can also be accomplished with a variety of other nucleic acid-binding dyes with affinity for DNA and resulting fluorescence greater than ethidium bromide (e.g., GelStar [Lonza], SYBR Green [Sigma]).
2. PFGE gel images may also by captured by photography (conventional, digital) although filters specific for the emitted fluorescent wavelength are required for optimal visualization. Image storage is less convenient than with more automated systems (e.g., Gel Doc [BioRad]).
3. A variety of size standard options exist including XbaI-restricted PulseNet global standard *Salmonella* Braenderup H2812 (ATCC BAA-664) used in all PulseNet protocols [6] and SmaI-digested *Staphylococcus aureus* NCTC 8325 (http://cmr.jcvi.org/cgi-bin/CMR/shared/MakeFrontPages.cgi?page=restriction_digest&crumbs=genomes). Inclusion of standards in at least the first and last gel lanes is important both for PFGE fragment size estimation as well as gel-to-gel normalization with computer analysis programs.
4. Alternatively, the gel can be precast and the agarose slices gently loaded into the wells. However, care must be taken not to fracture the slices during loading, and the wells should then be sealed with residual molten gel agarose to insure that the slices remain in the wells.

5. A variety of switching times may be used depending on the range of fragment sizes being separated [9]. The switching times presented here are from a harmonized protocol for MRSA but work equally well with *S. epidermidis.*
6. Disposal may differ depending on the stain used (*see* **Note 1**).
7. Interpreting PFGE data can be challenging especially in instances where the data is used to establish relationships between genotypes and epidemiologic events such as outbreaks. Nearly 20 years ago, Tenover et al. presented theoretical guidelines that remain frequently cited today regarding the process of PFGE analysis [7]. The approach proposed that a single genetic event in the chromosome of clinically relevant bacteria would be unlikely to represent an epidemiologically significant change especially in a short-term nosocomial environment (e.g., organism movement from patient to patient). Thus, isolates with indistinguishable PFGE patterns or differences due to a single genetic event were considered to represent the same epidemiological or strain type while isolates with additional differences were less likely to be related. However, as has been noted many times since [8, 9], the terminology was deliberately vague, acknowledging that the interpretation of molecular typing data by PFGE or any other method is dependent on a variety of factors especially including the stability of chromosomal sequences over time (i.e., the genomic clock speed). The impact of genetic events on chromosomal PFGE patterns has been extensively reviewed elsewhere [10, 11]. However, the simplest of genetic events (i.e., a single-base change) either creating or removing a restriction site results in a difference of three PFGE band positions (i.e., one band becoming two or two bands coalescing into one, respectively). Thus, a difference of no more than three restriction fragment positions has frequently been used as a "breakpoint" for relatedness with organisms of moderate genomic clock speed that would include *S. epidermidis.* In mathematical terms (depending on the total number of bands in the PFGE pattern) this generally equates to an association of isolate relatedness with ≥85 % pattern similarity; however, it must be emphasized that this is far from a hard-and-fast "rule." Additional caveats regarding PFGE interpretation include the following comments. First, genomic fragments differing up to ca. 10 % in size may not be reliably separated by PFGE and seen as distinct bands [10]. Second, depending on the switching parameters, chromosomal restriction fragments <30 kb in size are usually poorly separated and visualized in PFGE gels. Thus, pattern comparisons are only based on a (larger sized) subset of the restriction fragments generated. Third, the slower the genomic clock speed of the organism (degree of chromosomal change over time)

the more conservative one must be in defining a significant difference in the PFGE patterns of isolates (i.e., small differences become more important). Fourth, mobile genetic elements in the bacterial chromosome (e.g., transposons and plasmids) may also contribute to PFGE pattern variability as has been observed in *E. coli* O157 [12]. Fifth, no single guide or rule for interpreting PFGE typing data can be universally applied to all circumstances. Criteria for PFGE pattern analysis [9, 13] are a useful starting point but must be viewed as guidelines rather than rules which should not be used in isolation but as part of a broader effort of epidemiologic, environmental, and laboratory investigation.

References

1. Goering RV (2013) Molecular typing techniques: state of the art. In: Tang YW, Stratton CW (eds) Advanced techniques in diagnostic microbiology, 2nd edn. Springer, New York, NY, pp 239–261
2. Loman NJ, Misra RV, Dallman TJ et al (2012) Performance comparison of benchtop high-throughput sequencing platforms. Nat Biotechnol 30:434–439
3. Schatz MC, Langmead B, Salzberg SL (2010) Cloud computing and the DNA data race. Nat Biotechnol 28:691–693
4. Angiuoli SV, White JR, Matalka M et al (2011) Resources and costs for microbial sequence analysis evaluated using virtual machines and cloud computing. PLoS One 6:e26624
5. Murchan S, Kaufmann ME, Deplano A et al (2003) Harmonization of pulsed-field gel electrophoresis protocols for epidemiological typing of strains of methicillin-resistant *Staphylococcus aureus*: a single approach developed by consensus in 10 European laboratories and its application for tracing the spread of related strains. J Clin Microbiol 41:1574–1585
6. Hunter SB, Vauterin P, Lambert-Fair MA et al (2005) Establishment of a universal size standard strain for use with the PulseNet standardized pulsed-field gel electrophoresis protocols: converting the national databases to the new size standard. J Clin Microbiol 43:1045–1050
7. Tenover FC, Arbeit RD, Goering RV et al (1995) Interpreting chromosomal DNA restriction patterns produced by pulsed-field gel electrophoresis: criteria for bacterial strain typing. J Clin Microbiol 33:2233–2239
8. Goering RV (2000) The molecular epidemiology of nosocomial infection: past, present, and future. Rev Med Microbiol 11:145–152
9. Van Belkum A, Tassios PT, Dijkshoorn L et al (2007) Guidelines for the validation and application of typing methods for use in bacterial epidemiology. Clin Microbiol Infect 13(Suppl 3):1–46
10. Goering RV, Ribot EM, Gerner-Smidt P (2011) Pulsed-field gel electrophoresis: laboratory and epidemiologic considerations for interpretation of data. In: Persing DH, Tenover FC, Tang YW, Nolte FS, Hayden RT et al (eds) Molecular microbiology, 2nd edn. ASM Press, Washington, DC, pp 167–177
11. Goering RV (2010) Pulsed field gel electrophoresis: a review of application and interpretation in the molecular epidemiology of infectious disease. Infect Genet Evol 10:866–875
12. Bielaszewska M, Prager R, Zhang W et al (2006) Chromosomal dynamism in progeny of outbreak-related sorbitol-fermenting enterohemorrhagic Escherichia coli O157:NM. Appl Environ Microbiol 72:1900–1909
13. Goering RV, Tenover FC (1997) Epidemiological interpretation of chromosomal macro-restriction fragment patterns analyzed by pulsed-field gel electrophoresis. J Clin Microbiol 35:2432

Chapter 5

Multilocus Sequence Typing of *Staphylococcus epidermidis*

Jonathan C. Thomas and D. Ashley Robinson

Abstract

Multilocus sequence typing (MLST) is a genotyping method that is well suited for studying the population genetics and evolution of *Staphylococcus epidermidis*. The central MLST database for *S. epidermidis* continues to grow, and new analysis methods for extracting historical information from MLST data continue to be developed. Even in this era of whole-genome sequencing, MLST provides a reference genotyping method, and the central MLST database provides a unique catalog of genetic variants.

Key words *Staphylococcus epidermidis*, Multilocus sequence typing, Population biology, Molecular epidemiology, Population genetics, Evolution

1 Introduction

Knowledge of *Staphylococcus epidermidis* population genetics and evolution is currently quite limited [1–3]. The further development of this knowledge is reliant on an ability to identify and correctly interpret genetic variation. Classical methods for genotyping bacteria can indirectly identify nucleotide sequence variation and have been successfully applied to molecular epidemiological studies of *S. epidermidis*. These classical methods include pulsed-field gel electrophoresis (PFGE) [4, 5], amplified fragment length polymorphism (AFLP) [6], and multilocus variable number tandem repeat analysis (MLVA) [7], among others. Multilocus sequence typing (MLST) can directly identify nucleotide sequence variation and has several advantages for studies of population genetics and evolution [8]. Sequence data are unambiguous, easily shared between laboratories, and the subject of decades of development of theory about the origin and fate of genetic variation, which enables extensive modeling of sequence data.

MLST schemes typically target seven housekeeping genes, which are chosen for each bacterial species based on the genes' presence in all isolates of that species, separate locations on the

Paul D. Fey (ed.), *Staphylococcus Epidermidis: Methods and Protocols*, Methods in Molecular Biology, vol. 1106,
DOI 10.1007/978-1-62703-736-5_5, © Springer Science+Business Media, LLC 2014

bacterial chromosome, and functions that are not expected to be under strong diversifying selection. Internal fragments of each of the chosen housekeeping genes are sequenced for each isolate, and the sequences are compared against those in a central, curated MLST database. Each unique nucleotide sequence at a locus is assigned an allele. Each unique combination of alleles across the loci is assigned a sequence type (ST). Novel alleles and STs are added to the database, which provides a unique catalog of genetic variants. Three MLST schemes have been developed for *S. epidermidis* [9–11], but the scheme of Thomas et al. [11] provides a consensus of the other schemes and is the one scheme hosted at the central MLST database (http://sepidermidis.mlst.net).

MLST has been used mostly to unambiguously define bacterial clones at a coarse level of resolution (i.e., STs) and to infer genetic relationships between clones [1, 2]. Studies have also used MLST to measure and compare recombination rates in bacteria [12, 13] and to reveal unexpected signatures of positive selection within or near some MLST loci [3, 14, 15]. Methods for the analysis of MLST data continue to increase in sophistication and ability to extract historical information from the sequences. The importance of MLST as a reference genotyping method is highlighted by the recent development of software to retrieve MLST data from whole-genome sequences [16, 17]. The goal of this chapter is to describe the benchside methods used to generate MLST data in *S. epidermidis* and some analysis methods used to identify genetic clusters in this species.

2 Materials

2.1 Genomic DNA Extraction

1. Blood agar plates: Tryptic soy agar (TSA) plates, 5 % (v/v) sheep's blood.
2. Commercial DNA extraction kit.
3. Lysozyme.
4. Lysostaphin.

2.2 PCR Amplification of MLST Loci

1. 96-well PCR plate and adhesive PCR film.
2. Taq DNA polymerase (5 U/μL).
3. 10 mM deoxyribonucleotide triphosphates (dNTPs).
4. 25 mM $MgCl_2$.
5. PCR primers (10 pM) for the seven loci used in the *S. epidermidis* MLST scheme. The primers described by Thomas et al. [11] are used to amplify and sequence internal fragments of the following seven housekeeping genes: *arcC*, *aroE*, *gtr*, *mutS*, *pyrR*, *tpiA*, and *yqiL* (*see* Table 1).

Table 1
PCR and sequencing primers involved in the MLST of *S. epidermidis*

Locus	Primer name	Primer sequence (5′–3′)	Amplicon length (bp)	Trimmed allele length (bp)
arcC	arcC-F	TGTGATGAGCACGCTACCGTTAG	508	465
	arcC-R	TCCAAGTAAACCCATCGGTCTG		
aroE	aroE-F	CATTGGATTACCTCTTTGTTCAGC	459	420
	aroE-R	CAAGCGAAATCTGTTGGGG		
gtr	gtr-F	CAGCCAATTCTTTTATGACTTTT	508	438
	gtr-R	GTGATTAAAGGTATTGATTTGAAT		
mutS	mutS-F3	GATATAAGAATAAGGGTTGTGAA	608	412
	mutS-R3	GTAATCGTCTCAGTTATCATGTT		
pyrR	pyr-F2	GTTACTAATACTTTTGCTGTGTTT	851	428
	pyr-R4	GTAGAATGTAAAGAGACTAAAATGAA		
tpiA	tpi-F2	ATCCAATTAGACGCTTTAGTAAC	592	424
	tpi-R2	TTAATGATGCGCCACCTACA		
yqiL	yqiL-F2	CACGCATAGTATTAGCTGAAG	658	416
	yqiL-R2	CTAATGCCTTCATCTTGAGAAATAA		

6. Agarose gel and loading buffer.
7. 1× Tris–borate–EDTA (TBE) buffer, pH 8.3: 0.9 M Tris base, 0.9 M boric acid, 30 mM EDTA.
8. 10 mg/mL ethidium bromide.
9. 100 bp DNA ladder.
10. Sterile distilled water (SDW) (*see* **Note 1**).

2.3 Purification of PCR Amplicons and DNA Sequencing Reactions

1. Polyethylene glycol (PEG) solution: 20 % PEG-8000, 2.5 M NaCl mixture (w/v).
2. 95 and 70 % ethanol (v/v).
3. 3 M sodium acetate, pH 5.2.
4. 15 mL conical tube.

2.4 DNA Sequencing Reactions

1. 96-well PCR plate and adhesive PCR film.
2. BigDye® Terminator v3 cycle sequencing kit (Applied Biosystems). The version of BigDye® used in the sequencing reactions may differ by sequencing facility.
3. Sequencing primers (1 pM) for the seven loci used in the *S. epidermidis* MLST scheme (*see* **Note 2**).

3 Methods

3.1 Generation of MLST Data

3.1.1 Genomic DNA Extraction

1. Incubate bacterial isolates overnight at 37 °C on 5 % sheep's blood (v/v) TSA plates. Isolates should be streaked for single colonies.
2. Pick a single colony of each isolate, and streak for heavy growth on fresh 5 % sheep's blood (v/v) TSA plates.
3. Extract high-quality genomic DNA using a commercial kit, such as the DNeasy® kit (Qiagen), by following the manufacturer's instructions (*see* **Note 3**).
4. Store DNA at 4 °C.

3.1.2 PCR Amplification of MLST Loci (*See* Note 4)

1. Seven PCRs are performed for each *S. epidermidis* isolate, one PCR for each of the seven MLST housekeeping genes. Each PCR reaction comprises 2 μL of genomic DNA, 0.2 μL of Taq DNA polymerase, 5 μL of Taq DNA polymerase buffer, 2.5 μL of $MgCl_2$, 1 μL of dNTPs, 10.3 μL of SDW, and 2 μL (10 pM) of the appropriate forward and reverse PCR primers. PCR reactions should be prepared in 96-well PCR plates on ice and sealed with adhesive PCR film.
2. The PCR thermal cycling program consists of an initial denaturation step of 95 °C for 3 min, 34 cycles of 95 °C for 30 s, 50 °C for 1 min, and 72 °C for 1 min, followed by a final extension of 72 °C for 10 min and cooling to 4 °C.
3. To confirm whether the PCR was successful and that amplicons are of the appropriate size, run 5 μL samples of the PCRs on a 1 % agarose gel.

3.1.3 Purification of PCR Amplicons

1. Pipette 60 μL of the PEG/NaCl solution into each well of the PCR plate. Reseal the PCR plate with adhesive PCR film and vortex (*see* **Note 5**).
2. Centrifuge the PCR plate at 4 °C and 1,000 rpm for 10 s, and then incubate the plate at room temperature for 30 min.
3. Centrifuge the PCR plate at 4 °C and 2049 × *g* for 30 min.
4. Remove the adhesive PCR seal, and gently invert the PCR plate onto a paper tissue. Centrifuge the inverted plate at 4 °C and 500 rpm for 30 s to remove any excess PEG/NaCl solution. Repeat with a fresh paper tissue (*see* **Note 6**).
5. Wash the DNA pellet by adding 100 μL of chilled 70 % ethanol (v/v) per well. Reseal the PCR plate and centrifuge at 4 °C and 2049 × *g* for 30 min.
6. Remove the adhesive PCR seal, and gently invert the PCR plate onto a paper tissue to remove any excess ethanol.

Replace the paper tissue with a fresh paper tissue, and centrifuge the inverted plate at 4 °C and 500 rpm for 30 s.

7. Air-dry the PCR plate at room temperature for 10 min or at 37 °C for 2 min.
8. Add 12 μL of SDW into each well of the PCR plate. Reseal the PCR plate with a new adhesive PCR film.
9. Vortex the PCR plate for 1 min. Centrifuge at 4 °C and 1,000 rpm for 10 s.
10. Store the purified PCR amplicons at 4 °C. Ensure that the PCR plate is well sealed to avoid evaporation. For long-term storage, store the purified PCR amplicons at −20 °C.

3.1.4 DNA Sequencing Reactions

1. DNA sequencing reactions should be prepared in new 96-well PCR plates and consist of 2 μL of purified PCR amplicon, 1 μL of the forward sequencing primer for a given locus (1 pM; *see* **Note 2**), and 2 μL of BigDye® Terminators. Separate sequencing reactions should be prepared using the reverse sequencing primer.
2. Seal the PCR plate with adhesive PCR film and centrifuge at 4 °C and 1,000 rpm for 10 s.
3. The thermal cycling program for DNA sequencing reactions consists of 24 cycles of 95 °C for 10 s, 50 °C for 5 s, and 60 °C for 2 min. At the end of the 24 cycles, the temperature should decrease at 0.1 °C/s to 4 °C. Sequencing reactions are maintained at 4 °C in the thermal cycler until removed.

3.1.5 Purification of DNA Sequencing Reactions

1. Dilute each sequencing reaction by adding 16 μL of SDW.
2. Prepare an ethanol/sodium acetate solution by adding 6 mL of 95 % ethanol (v/v) and 240 μL of 3 M sodium acetate, pH 5.2, to a 15 mL conical tube. Vortex well to mix.
3. Pipette 52 μL of the 95 % ethanol/sodium acetate solution into each well of the PCR plate. Reseal the PCR plate with adhesive PCR film and vortex.
4. Centrifuge the PCR plate at 4 °C and 1,000 rpm for 10 s, and then incubate the plate at −20 °C for 30 min.
5. Centrifuge the PCR plate at 4 °C and 2049 × *g* for 30 min.
6. Remove the adhesive PCR film, and gently invert the PCR plate onto a paper tissue to remove any excess ethanol/sodium acetate solution. Replace the paper tissue with a fresh paper tissue, and centrifuge the inverted plate at 4 °C and 500 rpm for 30 s.
7. Wash the DNA pellet by adding 150 μL of 70 % ethanol (v/v) per well. Reseal the PCR plate before centrifuging at 2049 × *g* for 30 min.

8. Remove the adhesive PCR film, and gently invert the PCR plate onto a paper tissue to remove any excess ethanol. Replace the paper tissue with a fresh paper tissue, and centrifuge the inverted plate at 4 °C and 500 rpm for 30 s.
9. Air-dry the PCR plate at room temperature for 10 min. Reseal the PCR plate and store at −20 °C until it can be run on an automated DNA sequencer.

3.2 Analysis of MLST Data

3.2.1 Assigning Alleles and Sequence Types

1. The forward and reverse sequence trace files for each housekeeping gene of each isolate are assembled using a program such as SeqMan Pro, which is part of the Lasergene package (DNASTAR). A reference allele for each housekeeping gene may be downloaded from the central MLST database (http://sepidermidis.mlst.net) and used to trim the sequences to the appropriate start and stop points. Once the sequence trace files have been assembled, resolve any base-call differences between the two files. If a genuine conflict is observed between the two trace files, check that the appropriate trace files have been assembled; if the conflict remains, resequence the locus from a new PCR amplicon.
2. Alleles are assigned by querying the central MLST database. When alleles at all loci have been assigned with 100 % matches to alleles in the database, STs are assigned by again querying the database.
3. If novel alleles or combinations of alleles (i.e., STs) are discovered, communicate with the database curator to have new allele or ST numbers assigned. This involves submitting the trace files to the curator so that they may verify the novel sequences.

3.2.2 Assigning STs to Genetic Clusters

Genetic clusters of *S. epidermidis* may be inferred through various algorithms that use MLST data in the form of either alleles or nucleotide sequences. The eBURST algorithm [18] defines clusters based on the number of allelic differences between STs. This algorithm places the majority of all *S. epidermidis* STs into a single cluster and is therefore relatively uninformative about *S. epidermidis* population structure [19]. Furthermore, conventional phylogenetic algorithms that use nucleotide sequences and do not account for recombination events between STs may incorrectly cluster some STs. The Bayesian clustering algorithms BAPS [20] and STRUCTURE [21] use nucleotide sequences and, importantly, incorporate a model of population admixture. BAPS and STRUCTURE provide highly concordant results for clustering of *S. epidermidis* [22], but here we describe the use of BAPS as it has a shorter computation time.

1. Genetic clustering of *S. epidermidis* should be interpreted in the context of the global population structure. In practice, this

means that the MLST data of any study should be analyzed along with all *S. epidermidis* STs in the central MLST database (*see* **Note 7**).

2. Create a Microsoft Excel spreadsheet (.xls file extension). The first row is a header that includes ST in the first column and the names of the seven MLST loci in the subsequent seven columns (*see* **Note 8**). Subsequent rows include unique STs from your study and the STs from the remainder of the MLST database.
3. For each locus, trim and orientate the sequence so that it is in a +1 reading frame.
4. For each ST, enter the allele's sequence in its appropriate column.
5. Run the "Clustering with linked loci" analysis from BAPS v5, and select "MLST-format," followed by "Concatenate allelic sequences (Excel)."
6. Select the relevant file.
7. Save the pre-processed data.
8. Select the "Codon" linkage model.
9. Save the fully pre-processed data.
10. Input an upper bound to the number of populations. Multiple values may be entered at this point, and the analyses should be repeated with differing initial values, as recommended by the program's authors. An upper bound of 11 has proven sufficient for *S. epidermidis.*
11. Save the "Mixture populations" results file. This file will detail the number of clusters contained within the *S. epidermidis* population as well as the clusters to which the STs belong (*see* **Note 9**).
12. Select the "Admixture based on mixture clustering" analysis from BAPS v5.
13. Input the minimum size of a population to be used for the admixture analysis. A minimum of three individuals has been suggested by the program's authors.
14. Input the number of iterations to be run, followed by the number of reference individuals to be selected from each population and the number of iterations for each reference individual. Values of 100, 200, and 5–20, respectively, are suggested as appropriate initial values by the BAPS manual.
15. Save the "Admixture" results file. This file will detail any significantly admixed STs within the population, and it will describe the proportion of each ST's ancestry that belongs to each cluster.

4 Notes

1. All steps that require water should use SDW. Tap water may contain molecules that interfere with Sanger sequencing chemistry.
2. Sequencing primers are a 1:10 dilution of the primers used for PCR.
3. During the DNA extraction steps, the lysis of bacterial cells may be more effective with the addition of lysozyme (7.5 U/μL) and lysostaphin (0.075 U/μL).
4. The point at which the protocol should be concluded will be contingent on the template required by those performing the sequencing.
5. During the purification of PCR amplicons and DNA sequencing reactions, a new adhesive PCR film is not required at each step, but reseal the plate in the same orientation each time and avoid cross-well contamination. Steps that do require a new adhesive PCR film are indicated.
6. After the removal of excess PEG/NaCl solution from PCR plates by centrifugation, clean the inside of the centrifuge by wiping any excess PEG/NaCl not absorbed by the paper tissue. Failure to do so may lead to deterioration in the performance of the centrifuge.
7. Previous BAPS analysis of *S. epidermidis* MLST data indicates that correctly assigning strains to a cluster is strongly influenced by strain sampling [22].
8. Alternative data formats are available for the analysis of MLST data by BAPS, but we have found that the format described here is the simplest to assemble and maintain as the MLST database grows.
9. As of June 2012, the *S. epidermidis* MLST database consisted of 437 STs, from which six clusters are inferred using BAPS.

References

1. Kozitskaya S, Olson ME, Fey PD et al (2005) Clonal analysis of *Staphylococcus epidermidis* isolates carrying or lacking biofilm-mediating genes by multilocus sequence typing. J Clin Microbiol 43:4751–4757
2. Miragaia M, Thomas JC, Couto I et al (2007) Inferring a population structure for *Staphylococcus epidermidis* from multilocus sequence typing data. J Bacteriol 189:2540–2552
3. Zhang L, Thomas JC, Didelot X et al (2012) Molecular signatures identify a candidate target of balancing selection in an *arcD*-like gene of *Staphylococcus epidermidis*. J Mol Evol 75: 43–54
4. Kelly S, Collins J, Maguire M et al (2008) An outbreak of colonization with linezolid-resistant *Staphylococcus epidermidis* in an intensive therapy unit. J Antimicrob Chemother 61:901–907
5. Miragaia M, Carriço JA, Thomas JC et al (2008) Comparison of molecular typing methods for characterization of *Staphylococcus epidermidis*: proposal for clone definition. J Clin Microbiol 46:118–129

6. Sloos JH, Janssen P, van Boven CP et al (1998) AFLP typing of *Staphylococcus epidermidis* in multiple sequential blood cultures. Res Microbiol 149:221–228
7. Johansson A, Koskiniemi S, Gottfridsson P et al (2006) Multiple-locus variable-number tandem repeat analysis for typing of *Staphylococcus epidermidis*. J Clin Microbiol 44:260–265
8. Feil EJ, Enright MC (2004) Analyses of clonality and the evolution of bacterial pathogens. Curr Opin Microbiol 7:308–313
9. Wang XM, Noble L, Kreiswirth BN et al (2003) Evaluation of a multilocus sequence typing system for *Staphylococcus epidermidis*. J Med Microbiol 52:989–998
10. Wisplinghoff H, Rosato AE, Enright MC et al (2003) Related clones containing SCC*mec* type IV predominate among clinically significant *Staphylococcus epidermidis* isolates. Antimicrob Agents Chemother 47:3574–3579
11. Thomas JC, Vargas MR, Miragaia M et al (2007) Improved multilocus sequence typing scheme for *Staphylococcus epidermidis*. J Clin Microbiol 45:616–619
12. Pérez-Losada M, Browne EB, Madsen A et al (2006) Population genetics of microbial pathogens estimated from multilocus sequence typing (MLST) data. Infect Genet Evol 6:97–112
13. Vos M, Didelot X (2009) A comparison of homologous recombination rates in bacteria and archaea. ISME J 3:199–208
14. Zadoks RN, Schukken YH, Wiedmann M (2005) Multilocus sequence typing of *Streptococcus uberis* provides sensitive and epidemiologically relevant subtype information and reveals positive selection in the virulence gene *pauA*. J Clin Microbiol 43:2407–2417
15. Buckee CO, Jolley KA, Recker M et al (2008) Role of selection in the emergence of lineages and the evolution of virulence in *Neisseria meningitidis*. Proc Natl Acad Sci U S A 105: 15082–15087
16. Inouye M, Conway TC, Zobel J et al (2012) Short read sequence typing (SRST): multilocus sequence types from short reads. BMC Genomics 13:338
17. Larsen MV, Cosentino S, Rasmussen S et al (2012) Multilocus sequence typing of total-genome-sequenced bacteria. J Clin Microbiol 50:1355–1361
18. Feil EJ, Li BC, Aanensen DM et al (2004) eBURST: inferring patterns of evolutionary descent among clusters of related bacterial genotypes from multilocus sequence typing data. J Bacteriol 186:1518–1530
19. Smyth DS, Robinson DA (2010) Population genetics of staphylococcus. In: Robinson DA, Falush D, Feil EJ (eds) Bacterial population genetics in infectious disease, 1st edn. Wiley-Blackwell, NJ
20. Corander J, Marttinen P, Siren J et al (2008) Enhanced Bayesian modeling in BAPS software for learning genetic structures of populations. BMC Bioinformatics 9:539
21. Falush D, Stephens M, Pritchard JK (2003) Inference of population structure using multilocus genotype data; linked loci and correlated allele frequencies. Genetics 164: 1567–1587
22. Thomas JC, Zhang L, Robinson DA (2013) Differing lifestyles of *Staphylococcus epidermidis* as revealed through Bayesian clustering of multilocus sequence types. Infect Genet Evol 10.1016/j.meegid.2013.06.020

Chapter 6

Growth and Preparation of *Staphylococcus epidermidis* for NMR Metabolomic Analysis

Greg A. Somerville and Robert Powers

Abstract

The "omics" era began with transcriptomics and this progressed into proteomics. While useful, these approaches provide only circumstantial information about carbon flow, metabolic status, redox poise, etc. To more directly address these metabolic concerns, researchers have turned to the emerging field of metabolomics. In our laboratories, we frequently use NMR metabolomics to acquire a snapshot of bacterial metabolomes during stressful or transition events. Irrespective of the "omics" method of choice, the experimental outcome depends on the proper cultivation and preparation of bacterial samples. In addition, the integration of these large datasets requires that these cultivation conditions be clearly defined [1].

Key words Bacterial growth, Growth conditions, Metabolism, Metabolomics, NMR

1 Introduction

Staphylococci live in, and transition between, diverse host environments that can lead to rapid changes in the availability of nutrients necessary to provide energy and biosynthetic intermediates for the synthesis of macromolecules. These changes in nutrient availability alter the bacterial metabolic status, which cause regulatory changes that help bacteria adapt and survive in the new environment [2]. Changes in nutrient availability not only occur when bacteria are growing on, or in, a host environment, but they also occur during non-steady-state cultivation. Since most bacterial cultivation in microbiology laboratories is done using batch cultures, an understanding of the in vitro cultivation of *Staphylococcus epidermidis* is essential for experimental success. Achieving the reproducible cultivation of *S. epidermidis*, or any bacterium, requires consideration of three things: bacterial strain, culture medium, and cultivation conditions [3].

Paul D. Fey (ed.), *Staphylococcus Epidermidis: Methods and Protocols*, Methods in Molecular Biology, vol. 1106, DOI 10.1007/978-1-62703-736-5_6,

1.1 Bacterial Strain Selection

There are two schools of thought regarding strain selection: the first advocates a homogeneous approach using standard strains, that is to say, have many laboratories studying one common strain in great detail to maximize understanding of that particular strain. The second school of thought uses a heterogeneous approach with multiple strains to gain a broader understanding of the bacterial species population. We will leave it to the readers to decide which approach is more valuable; however, we would note that given the large number of *S. epidermidis* strains in use around the world it seems that most researchers have decided on the heterogeneous approach.

1.2 Culture Medium

There are three types of culture media: chemically defined, complex, and selective. A chemically defined medium (CDM) is one whose exact chemical composition is known [4]. Complex media are composed of digests of chemically undefined substances such as yeast, soy, or meat extracts, for example, casamino acids–yeast extract–glycerophosphate (CYGP) broth [5]. Finally, selective media can be made from either chemically defined or complex media and are used to enhance the isolation of particular bacterial species (e.g., mannitol salt agar). In NMR metabolomic studies, the choice of medium will largely depend on which isotopically labeled metabolite is being followed. For example, when using ^{15}N-glutamate, it is impractical to add labeled glutamate to a complex medium containing an unknown concentration of unlabeled glutamate. In this example, to achieve maximal labeling of the bacteria, it would be best to use a CDM lacking glutamate and glutamine. In our laboratories, we typically label staphylococci using ^{13}C-glucose in the complex medium tryptic soy broth (TSB) that is devoid of unlabeled glucose [6, 7]. This medium allows for maximal biomass generation while assuring that nearly all (~99 %; 1.1 % is due to naturally occurring ^{13}C) of the ^{13}C-labeled metabolites in the metabolome were derived from glucose.

1.3 Cultivation of *S. epidermidis*

In order to integrate the large quantity of genomic, proteomic, and metabolomic data that is rapidly accumulating in the literature, there must be a common thread between these studies that permits the direct comparison of data. This common thread must start with clearly defined growth conditions [1]. Unfortunately, as microbiology has progressed into the "omics" age, little emphasis has been placed on the necessary rigor needed to cultivate bacteria in a reproducible manner that permits the integration of data from different laboratories. The importance of having clearly defined growth conditions is easily demonstrated using the example of staphylococcal pyruvate catabolism [8]. During microaerobic or anaerobic growth, pyruvate is primarily reduced to lactic acid with the concomitant oxidation of NADH to NAD$^+$. In contrast, during aerobic growth pyruvate undergoes oxidative decarboxylation to produce acetyl-CoA, which is used to synthesize the small

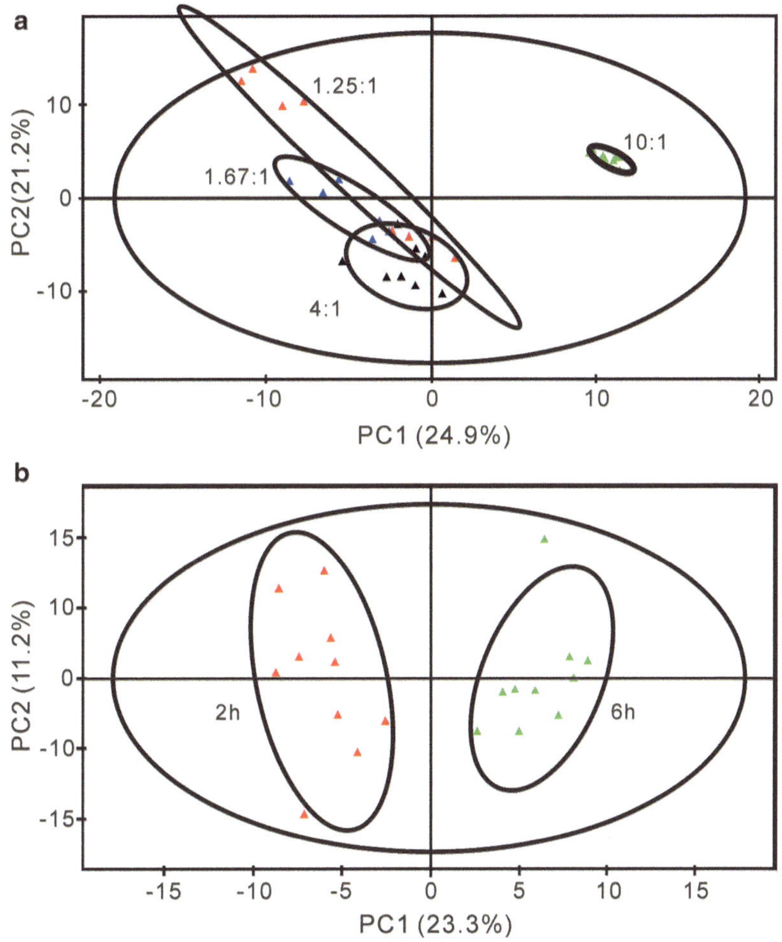

Fig. 1 Staphylococcal metabolism dramatically changes in response to changes in the flask-to-medium ratio and during the transition from the exponential phase of growth to the post-exponential growth phase. (**a**) Principal component analysis (PCA) of *S. aureus* metabolomes grown with differing flask-to-medium ratios (*indicated in the figure*). (**b**) PCA scores plot of the metabolomes of exponential (2 h) and post-exponential (6 h) growth phases for *S. epidermidis* strain 1457 grown at 37 °C with a flask-to-medium ratio of 10:1 and 225 rpm aeration. The ellipses correspond to the 95 % confidence limits from a normal distribution for each cluster

phospho-donor acetyl-phosphate. Acetyl-phosphate serves as the substrate for acetate kinase in substrate-level phosphorylation to generate ATP and acetic acid. It is obvious from this example that the metabolomes of staphylococci grown under microaerobic, anaerobic, or aerobic conditions would be different, hence the need to clearly define the growth conditions (Fig. 1a). At a minimum, microbiologists must define the cultivation medium, temperature, pH buffering (if used), % CO_2 (if used), flask-to-medium ratio, revolutions per minute of agitation (if used), use of baffled or non-baffled flasks, and inoculum dose. Ideally, it would be best to report the growth rates for every set of cultivation conditions [1].

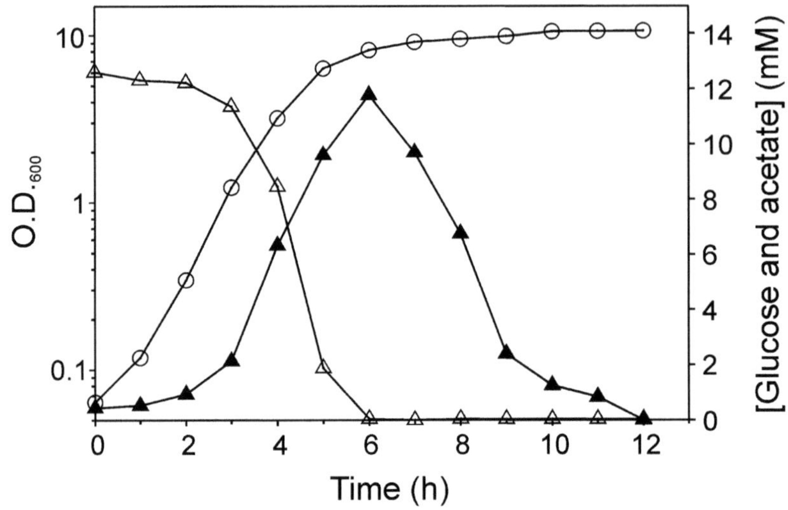

Fig. 2 Growth analysis of *S. epidermidis* showing the temporal depletion of glucose and the accumulation and depletion of acetate. *S. epidermidis* strain 1457 was grown in TSB at 37 °C with a flask-to-medium ratio of 10:1 and 225 rpm aeration

While reporting the cultivation conditions essential for integrating "omics" data, it is equally important to be aware that staphylococcal metabolism also changes during growth-phase transitions and that these changes can have regulatory consequences [2, 8].

1.4 Growth Phase-Dependent Differences in Metabolism

A non-steady-state bacterial growth profile consists of lag, exponential, stationary, and death phases. The lag phase is the time during which bacteria are increasing the cell size and mass in preparation for the rapid growth to follow. After the bacteria have achieved their initiation mass, they enter into a period of exponential, not logarithmic, growth. Exit from the exponential growth phase usually occurs when one or more nutrients are depleted from the medium or when there is an accumulation of a toxic by-product of metabolism. Following the exponential growth phase, the bacteria enter into the stationary phase, which is characterized by slower, or no, growth. Although not technically defined as a growth phase, the period after the end of the exponential growth phase, but prior to the cessation of growth, is often referred to as the post-exponential growth phase. In staphylococci, the exponential and post-exponential growth phases coincide with two very distinct metabolic states (Figs. 1b and 2).

As stated above, for NMR metabolomic studies, we aerobically cultivate *S. epidermidis* in TSB containing uniformly labeled ^{13}C-glucose. In the exponential growth phase, glucose is primarily catabolized through the glycolytic and pentose phosphate pathways. This is due in part to the CcpA-dependent repression of tricarboxylic acid (TCA) cycle activity, which prevents the flow of carbon into the TCA cycle [9]. The aerobic growth conditions together with a minimally active TCA cycle shunt the pyruvate from glycolysis through the pyruvate dehydrogenase complex to generate acetyl-CoA, which is used in the phosphotransacetylase/acetate kinase pathway to produce acetyl-phosphate [8]. Acetyl-phosphate is used for substrate-level phosphorylation to generate ATP, but it also leads to the accumulation of acetate in the culture medium (Fig. 2). In contrast to aerobic growth conditions that we typically employ, the heterofermentative nature of *S. epidermidis* will lead to the accumulation of lactic acid and small amounts of formate, ethanol, acetate, and other alcohols during microaerobic or anaerobic growth. This aerobic/anaerobic change in catabolic profiles is similar to Louis Pasteur's observation that in the absence of oxygen yeasts consume more glucose than in the presence of oxygen and that this increased consumption coincided with alcoholic and lactic acid fermentation [10]. In the laboratory, the transition from aerobic growth to fermentative growth can occur by simply decreasing the flask-to-medium ratio from 10:1 to 4:1 (Fig. 1a); for this reason, it is critical to state the exact growth conditions.

Whether *S. epidermidis* is aerobically or anaerobically cultivated, the exponential growth phase is characterized by the rapid catabolism of carbohydrates and the accumulation of partially oxidized acidic and alcohol by-products of pyruvate catabolism (Fig. 2). In the absence of oxygen or other electron acceptor(s) (e.g., nitrate), growth will cease shortly after the depletion of carbohydrates due to an inability to complete the oxidation of these by-products. When oxygen is available, the bacteria will catabolize the acidic and alcohol by-products, a process requiring the TCA cycle. This second distinct metabolic state coincides with the post-exponential growth-phase derepression of the TCA cycle, gluconeogenesis, and numerous biosynthetic pathways. Using our example of aerobically cultivated *S. epidermidis*, acetate is taken up from the culture medium (Fig. 2) and enters the TCA cycle in the form of acetyl-CoA via a condensation reaction with oxaloacetate to generate citric acid. The complete oxidation of acetate by way of the TCA cycle creates biosynthetic intermediates and reducing potential to drive biosynthesis and oxidative phosphorylation, which supports post-exponential growth in a medium depleted of readily catabolizable carbohydrates. In summary, the transition between the two distinct staphylococcal metabolic states can be followed by monitoring the bacterial growth phases.

1.5 Spectrophotometric Evaluation of Bacterial Growth

When cultivating staphylococci for metabolomic analysis, the monitoring of bacterial growth is a critical parameter due to the fact that variation in growth conditions, growth phases, or cell numbers will be reflected in the NMR spectra. To assess bacterial growth, there are three commonly used practices: spectrophotometry, microscopic counting, and determination of colony-forming units (cfu)/mL [11]. For a rapidly growing bacterium, the determination of cfu in real time is not possible; hence, this method is better suited for confirming the number of cfu taken at harvest than it is for determining how many mL of culture must be harvested to achieve equivalent bacterial numbers (*see* **Note 1**). Similarly, microscopic counting of bacterial numbers (e.g., using a Petroff-Hausser counting chamber) is impractical for staphylococci due to the fact that they often form irreducible clusters. This leaves spectrophotometry as the method of choice for monitoring the real-time growth of staphylococci.

In principle, spectrophotometers pass a monochromatic beam of light, with an intensity of I_o, through a gas, solution, or solid having a defined path length (l) and then record the intensity of the light (I) that emerges from the sample. Under these conditions, transmission (T) = I/I_o and the attenuation of the beam are dependent upon two independent properties: the frequency of the light and the nature of the sample. When monitoring bacterial growth using a spectrophotometer, the nature of the sample is such that there is minimal absorbance of light, but there is scattering of the light. The net result of light scattering in a spectrophotometer is that the intensity of the beam is attenuated due to the redirection of light away from the detector, not due to being absorbed by the sample. To accurately reflect the beam attenuation, it is essential to adequately dilute the sample so as to avoid multiple scattering events. The degree to which a sample must be diluted is dependent upon the wavelength and the size and shape of the scattering particles. In addition, instrument limitations must be taken into account so as to maintain the concentration (turbidity) within the linear range of the spectrophotometer. For each spectrophotometer and wavelength, it is appropriate to empirically determine the dilutions necessary to maintain the absorbance (aka, optical density; O.D.) within the linear range of the spectrophotometer. In our laboratories, we typically measure the O.D. using a 600 nm wavelength in a 1 cm path length cuvette. As the culture turbidity approaches the maximum linear range of a spectrophotometer (this is usually between 0.8 and 1.0 $O.D._{600}$ units), the samples are diluted into water or culture medium. Typically, these dilutions are between 1:5 and 1:50 and are done to maintain the $O.D._{600}$ below about 0.6. In rich culture media, the optical density at a wavelength of 600 nm for a culture of *S. epidermidis* will rarely achieve 20 $O.D._{600}$ units; hence, it is unlikely that dilutions greater than 1:50 will be necessary.

1.6 Sample Preparation

NMR metabolomic sample preparation requires three things: speed, quantity, and consistency. *Speed* is required to capture an accurate "snapshot" of the metabolome, particularly, unstable metabolites, such as those having high energy phosphate bonds (e.g., acetyl-phosphate) may experience concentration changes if the sample preparation time is too long. The greater the *quantity* of metabolites harvested, the stronger the NMR signal, and the greater the likelihood that metabolites of lower abundance will be identified. *Consistency* is essential due to the sensitivity of the NMR to artificial variation. Inconsistency in sample preparation may cause artificial differences that will be reflected in statistical analyses (e.g., principal component analysis (PCA)) or that may obscure small but significant metabolic differences. While there are numerous methods of preparing samples for NMR metabolomic analysis, a number of recent reviews on these methods are available [12, 13]; therefore, we will only describe the technique most commonly used in our laboratories in Subheading 3.

1.7 NMR Data Collection

There are two general goals of an NMR metabolomics study: The first is to capture the global state of the metabolome (metabolic fingerprinting), which is typically accomplished using a one-dimensional (1D) ^{1}H NMR experiment. The second primary goal is to identify the metabolites that differentiate two or more bacteria or bacterial growth conditions, specifically, which metabolites are undergoing a significant concentration change or are completely missing in a particular bacterial type or culture condition. In our laboratories, this is typically accomplished by using two-dimensional (2D) ^{1}H–^{13}C heteronuclear single-quantum coherence (HSQC) NMR experiments. In the following sections, we discuss some important considerations for each type of experiment.

1.7.1 1D ^{1}H NMR

The 1D ^{1}H NMR spectrum contains hundreds to thousands of spectral lines or peaks, where each peak arises from a specific hydrogen nucleus (or proton in NMR parlance) from each metabolite, buffer, or solvent molecule present in the sample (Fig. 3). To minimize obscuring metabolite peaks by buffer or solvent peaks, a PBS buffer made with deuterium oxide (D_2O) is used as the solvent for the NMR metabolomic samples (*see* **Note 2**). Despite using D_2O, there are still some residual H_2O peaks that need to be removed using solvent suppression. To achieve this, we incorporate excitation sculpting into our 1D ^{1}H pulse sequence to remove the water signal and maintain a flat spectral baseline [14]. A common alternative approach employs a 1D ^{1}H NOESY experiment, described in [15]. In addition to water, the presence of proteins or other biomolecules in the sample may also obscure the metabolite peaks. Large-molecular-weight proteins usually cause an underlying broadness to the NMR spectrum and a distorted “smile” shape to the baseline. Any protein contamination can be chemically removed by precipitation with methanol or other equivalent organic solvents; however, this may result in the unintended loss of metabolites. Alternatively, the NMR pulse sequence can be modified by using a Carr–Purcell–Meiboom–Gill (CPMG) spin-echo system [16]. Simply, the CPMG pulse sequence selects for metabolites over protein signals based on the large-molecular-weight difference.

After the removal of water and/or protein peaks, metabolite peaks are assigned chemical shifts that typically range from 0 to 10.0 ppm, which depends on the unique chemical environment of each proton. Maintaining a consistent chemical shift assignment between multiple

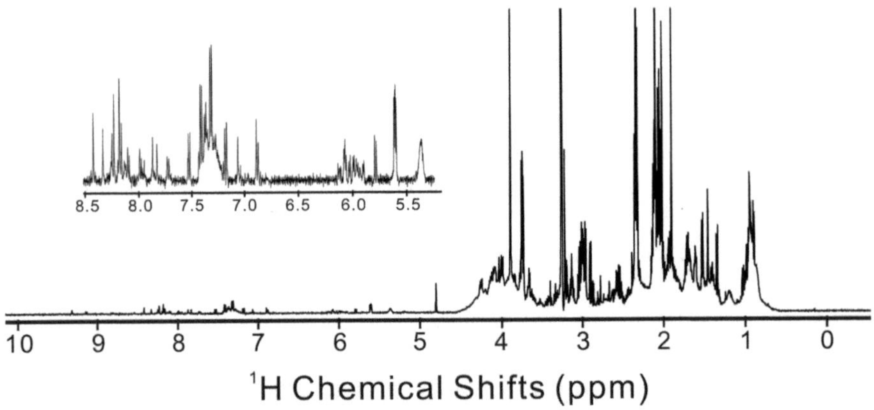

Fig. 3 Illustration of a typical one-dimensional ^{1}H NMR spectrum of an *S. epidermidis* strain 1457 cell-free lysate. The *inset* is an enhanced view of the aromatic region where the peak intensity has been increased by a factor of 12

samples requires using an internal standard. In our laboratories, we add 50 μM 3-(trimethylsilyl)propionic acid-2,2,3,3-d4 (TMSP-d4) to the NMR buffer and assign the single methyl peak to 0 ppm. To further minimize any instrument variation and bias, the NMR spectra are collected in an automated fashion using a robotic sample changer and software that automates sample tuning, matching, shimming and locking, and data collection. Furthermore, replicate samples from each class are randomly interleaved to avoid the introduction of any unintended bias into the spectral data.

1.7.2 Two-Dimensional ^{1}H–^{13}C HSQC

As stated above, we routinely use (2D) ^{1}H–^{13}C HSQC to identify metabolites and to determine the concentrations of those metabolites. Alternatively, 2D ^{1}H-^{1}H total correlation spectroscopy (TOCSY) [17] experiments, and statistical methods such as statistical total correlation spectroscopy (STOCSY) [18] are commonly used for metabolite identification (Fig. 4). The 2D ^{1}H–^{13}C HSQC NMR experiment generally requires including a ^{13}C-labeled metabolite in the culture medium; hence, 2D ^{1}H–^{13}C HSQC NMR monitors the flow of carbon-13 through the metabolome, and only metabolites derived from the ^{13}C-labeled metabolite will be observed in the 2D ^{1}H–^{13}C HSQC spectrum. Since 2D ^{1}H–^{13}C HSQC experiments may require an hour or longer to acquire a spectrum compared to approximately 5 min for 1D ^{1}H NMR experiments, the spectrum is only collected in triplicate. To minimize data acquisition time in 2D NMR experiments, sample sensitivity can be increased by increasing the number of bacteria harvested.

1.8 NMR Data Analysis

An NMR metabolomics study attempts to understand the changes in the cellular metabolome resulting from environmental stress, drug treatment, different bacterial strains or mutations, or any number of experimental conditions or biological systems. The NMR spectrum captures the state of the metabolome in order to make

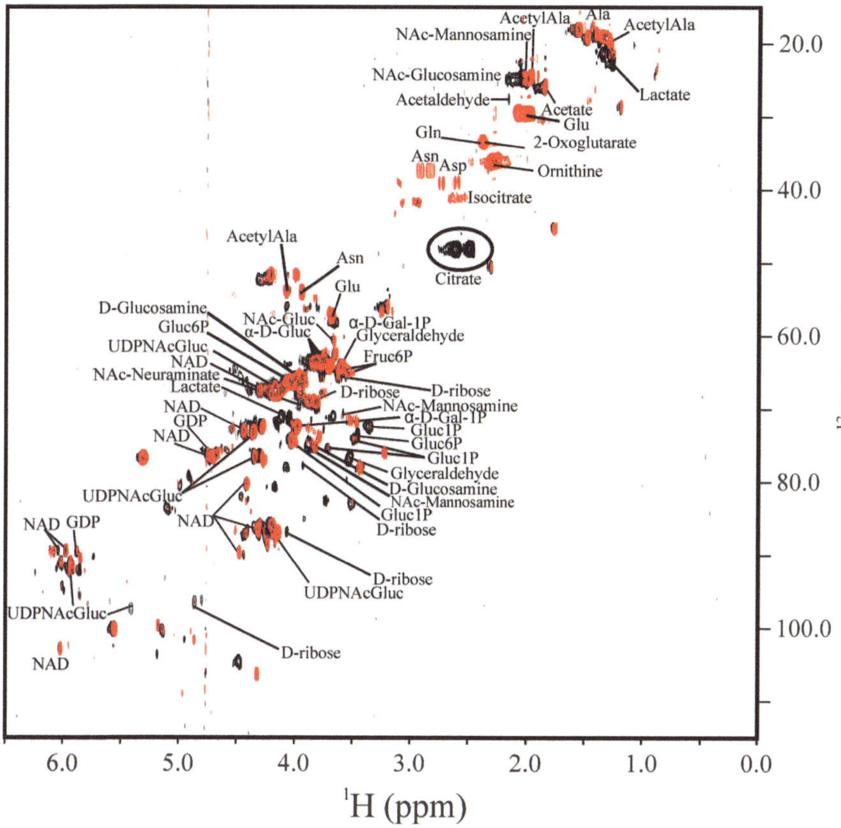

Fig. 4 Overlay of 2D ^{1}H–^{13}C HSQC spectra comparing *S. epidermidis* strain 1457 (*red*) and an isogenic aconitase mutant 1457-*acnA*::*tet*M (*black*) grown for 6 h in TSB medium containing 0.25 % ^{13}C-glucose. NMR resonances corresponding to specific metabolites are labeled, where citrate is *circled*. Reprinted with permission from Zhang B, Halouska S, Schiaffo CE, Sadykov MR, Somerville GA, Powers R: NMR analysis of a stress response metabolic signaling network. J Proteome Res 2011, 10:3743–3754 [49]. Copyright 2011 American Chemical Society

these comparisons. Thus, it is essential that the observed changes in the NMR spectrum are biologically relevant and not a result of sample preparation, data collection, or data processing. Correspondingly, consistency is critical to a metabolomics study. All sample preparation and NMR data collection should follow the identical protocol, use the same equipment and chemicals, and be conducted by the same individual at the same time, within reason. Similarly, uniform pre-processing of the NMR data is essential in order to obtain reliable and interpretable results from multivariate analysis techniques.

Pre-processing of the NMR spectra consists of (1) binning or aligning, (2) normalization, (3) scaling, (4) and noise removal. These issues have been previously discussed in detail (*see* refs. 19–22) and are only briefly summarized here. Chemical shifts or the spectral lines in an NMR spectrum are extremely sensitive to experimental conditions. As a result, it is critical that multiple NMR spectra are properly aligned so we are monitoring changes in the same peaks across the entire set of NMR spectra. The most common approach, and the method we employ, to solving this issue is the use of "adaptive" or "intelligent" binning [23–26]. The NMR spectrum is divided into a series of bins with widths ranging from 0.025 to 0.040 ppm. The bin edges are selected in a manner to avoid splitting peaks between bins. While binning ensures that we are comparing identical peaks between spectra, each spectrum must also be normalized to prevent the introduction of false concentration differences. To normalize each NMR spectrum, we use bacterial growth (*see* Subheading 1.5) to confirm that each sample has the same number of bacteria. After global normalization, there is still considerable variability in NMR peak intensities, even within a group of replicate samples, due to experimental variation. We address this problem by center averaging (also known as standard normal variate (SVN) or a Z-score) each spectrum that emphasizes correlations [22]:

$$Z = \frac{x_i - \bar{x}}{\sigma}$$

where $\bar{x}$ is the average signal intensity, σ is the standard deviation in the signal intensity, and x_i is the signal intensity within the bin. The next step of the data pre-processing is to remove noise, solvent, or buffer regions [20]. Again, normalization removes biologically irrelevant variations between samples due to differences in the total sample concentration.

The pre-processed NMR data is then interpreted using multivariate statistical techniques. We primarily use PCA and orthogonal partial least-squares discriminant analysis (OPLS-DA) [27–29]. OPLS-DA and PCA are performed using SIMCA-P11.5+ (UMETRICS), which automates scaling of the bins (columns) between each sample (row). SIMCA-P uses unit variance by default, but there are a variety of scaling protocols available depending on the specific goals of the metabolomics study. Scaling diminishes irrelevant contributions of intense peaks to the multivariate statistical analysis. Simply, relatively small percentage changes in intense peaks would dominate large percentage changes in weak peaks in the absence of scaling. This occurs because PCA and OPLS-DA simply emphasize absolute changes in peak intensities. The primary outcome of PCA and OPLS-DA is a scores plot, where each NMR spectrum is represented as a single point in PC space (Fig. 5). Briefly, each NMR spectrum is reduced to a single point in multidimensional space, where each axis is a bin and the value along the axis is the bin integral. Each NMR spectrum is then fitted to a vector $\left(\overrightarrow{PC_1}\right)$ corresponding to the largest variation in the data and then to a second vector orthogonal $\left(\overrightarrow{PC_2}\right)$ to the first. The 2D scores plot represents the projection of each spectra into the hyperplane described by $\left(\overrightarrow{PC_1}\right)$ and $\left(\overrightarrow{PC_2}\right)$. The relative clustering in the resulting scores plot identifies the relative similarities or differences between the 1D ^{1}H NMR spectrum and, correspondingly, the metabolomes. Thus, biological significance is

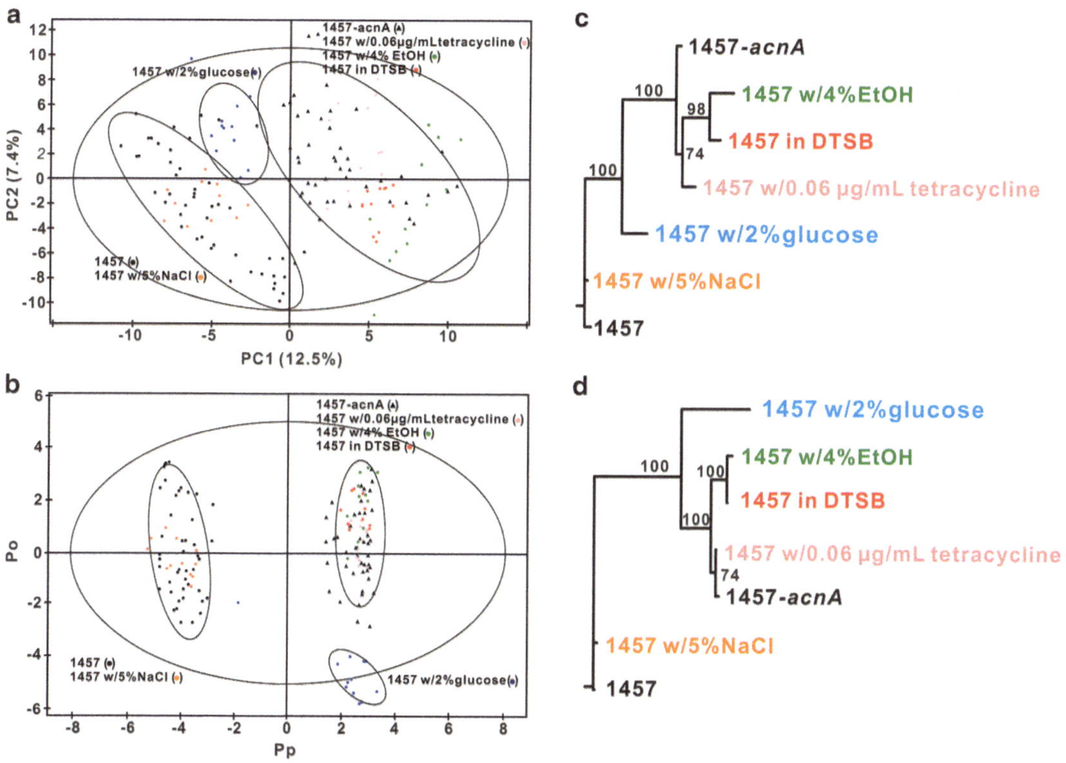

Fig. 5 Statistical analysis of *S. epidermidis* metabolomes prepared from stressed cultures aids in assessing the relationship between the stressors. (**a**) 2D PCA scores plot and (**b**) 2D OPLS-DA scores plot comparing *S. epidermidis* strain 1457 grown for 6 h in TSB medium (*black-filled circle*), with *S. epidermidis* 1457 cells grown 6 h in iron-depleted media (DTSB) (*red-filled circle*), with the addition of 4 % ethanol (*green-filled circle*),with the addition of 2 % glucose (*blue-filled circle*), with the addition of 0.06 μg/mL tetracycline (*pink-filled circle*), and with the addition of 5 % NaCl (*orange-filled circle*), and 6-h growth of aconitase mutant strain 1457-*acnA*::*tet*M in standard TSB medium (*black-filled triangle*). The ellipses correspond to the 95 % confidence limits from a normal distribution for each cluster. For the OPLS-DA scores plot, the 6-h growth of wild-type *S. epidermidis* 1457 (*black-filled circle*) was designated the control class and the remainder of the cultures were designated as treated. The OPLS-DA used 1 predictive component and 4 orthogonal components to yield a R^2X of 0.637, R^2Y of 0.966, and Q^2 of 0.941. CV-ANOVA validation yielded an effective *p*-value of 0.0. Metabolomic tree diagram generated from the (**c**) 2D PCA scores plot depicted in (**a**) and (**d**) 2D OPLS-DA scores plot depicted in (**b**). The *label colors* match the *symbol colors* from the 2D scores plot. Each node is labeled with the boot-strap number, where a value above 50 indicates a statistically significant separation. Reprinted with permission from [49], copyright 2011 American Chemical Society

inferred by the relative clustering or class discrimination in the PCA or the OPLS-DA scores plot. To infer statistical relevance to the clustering pattern, we use both an ellipse that defines the 95 % confidence limit from a normal distribution for each class and hierarchical clustering similar in concept to phylogenetic tree diagrams (Fig. 5) [30, 31]. The hierarchal clustering is based on the distance in the scores plot between each defined cluster or class. An updated version of our PCAtoTree software package is distributed under version 3.0 of the GNU General Public License and is freely available at http://bionmr.unl.edu/pca-utils.php.

In contrast to PCA, OPLS-DA is a supervised technique in which each NMR spectrum is typically assigned to one of the two binary classes, control (0) and treated (1). The interpretation is thus "biased" by the class definition and requires the OPLS-DA model to be validated. Essentially, OPLS-DA will identify class separation even for random data [32]. The identification of a similar class separation between the PCA and OPLS-DA models supports the reliability of the OPLS-DA model, but it is not sufficient for model validation. We use the leave-*n*-out method [33, 34] that reports quality assessment (Q^2) and quality of fit (R^2) values. These statistics simply reflect the consistency between the data and the model. A Q^2 value of ≥ 0.4 infers an acceptable model for a biological system [35], but large Q^2 value can also result from invalid or irrelevant models; therefore, we also use CV-ANOVA [36] to further validate the model with a standard *p*-value.

After identifying a class separation based on the PCA or the OPLS-DA scores plot, the next goal is to identify which metabolites primarily contribute to the class distinction. One common approach we use is to identify significant metabolite concentration changes by using the time-zero HSQC experiment [37]. The 2D ^1H–^{13}C HSQC NMR correlates carbon and hydrogen chemical shifts for every bonded C–H pair. The higher spectral resolution (2D vs. 1D) and the multiple chemical shift information (^1H and ^{13}C) significantly improve the reliability of identifying metabolites. Simply, the experimental chemical shifts obtained from the 2D ^1H–^{13}C HSQC spectrum are matched against reference spectral data in databases such as the Madison Metabolomics Consortium Database [38], the BioMagResBank [39], and the Human Metabolome Database (HMDB) [40]. There are two practical challenges: (1) Metabolomics is a relatively new field, so the databases are incomplete, and (2) ^1H and ^{13}C chemical shift tolerances of 0.05 and 0.50 ppm, respectively, are used to identify a match between the experimental and database chemical shifts because of experimental variation, potentially leading to multiple possible matches. For these reasons, it is important to incorporate biological information to refine the metabolite assignments.

Other important outcomes of OPLS-DA are an S-plot and loadings plot (Fig. 6). An S-plot identifies the chemical shift

Fig. 6 (continued) Each point was identified to a specific metabolite using the Human Metabolomics Database and Madison Metabolomics Database. All the identified metabolites are associated with TCA cycle inactivation. (**b**) OPLS-DA S-plot comparing the mutant strain 1457-*acnA*::*tetM* grown for 2 and 6 h. The metabolites identified are associated with variations in the utilization of glucose. (**c**) OPLS-DA loading plot comparing *S. epidermidis* 1457 and aconitase mutant strain 1457-*acnA*::*tetM* where both cell cultures were grown for 6 h. Negative values indicate a decrease in peak intensity when comparing the wild-type to the aconitase mutant, while positive values indicate an increase in peak intensity. (**d**) OPLS-DA loading plot comparing the mutant strain 1457-*acnA*::*tetM* grown for 2 and 6 h. Reprinted with permission from [49], copyright 2011 American Chemical Society

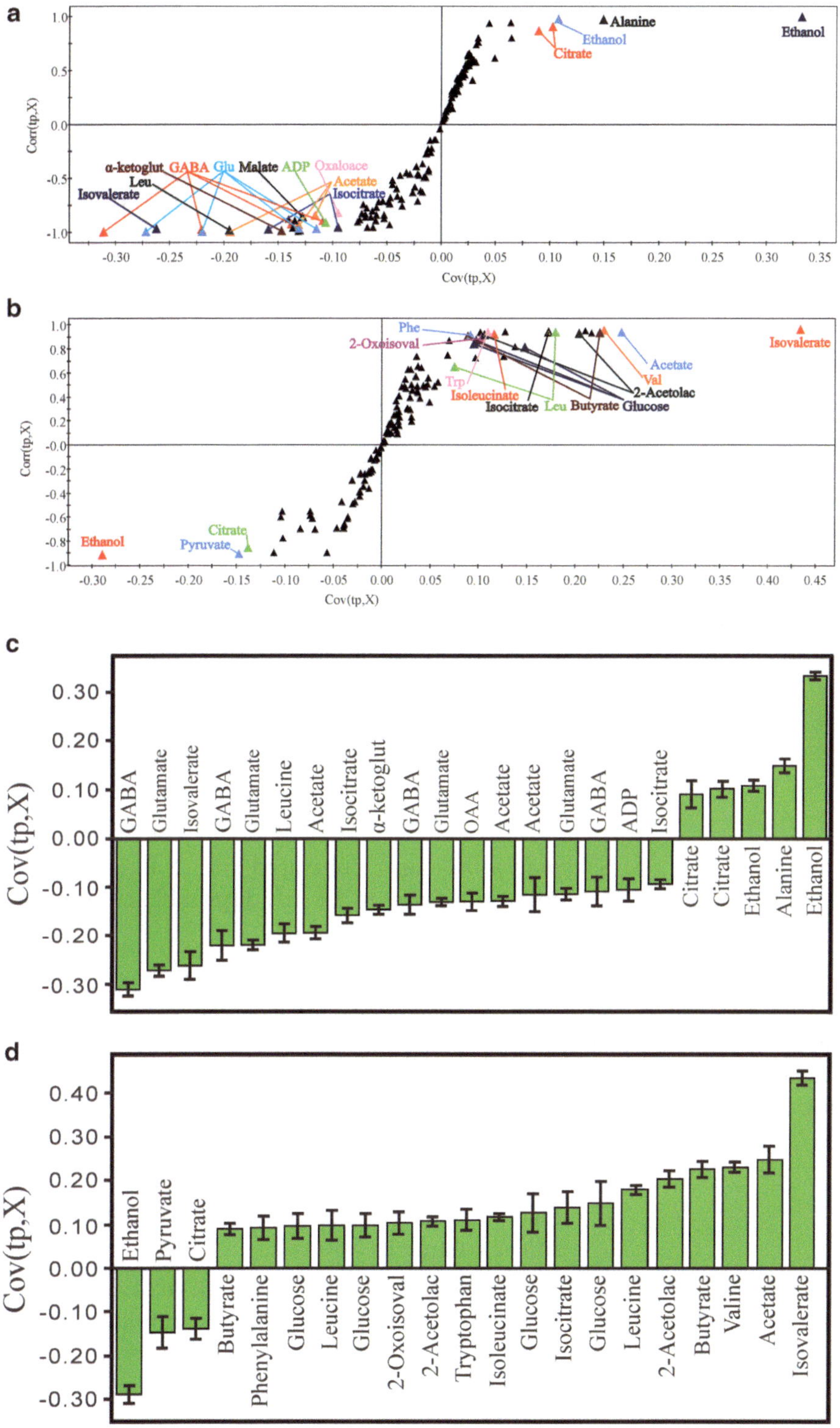

Fig. 6 (**a**) OPLS-DA S-plot comparing the post-exponential (6 h) metabolomes of *S. epidermidis* strain 1457 and its isogenic aconitase mutant (1457-*acnA*::*tetM*). The metabolomes were shown to be separated along the PC1 axis in Fig. 5a. Each point in the S-plot represents a specific bin containing a chemical shift range of 0.025 ppm, where the points at the extreme ends of the S-plot are the major contributors to the class distinction.

bins and, correspondingly, the metabolites that are the major contributors to the class discrimination. Similarly, the loadings plot identifies the magnitude of each bin's contribution to class discrimination, which can be interpreted as a relative measure of metabolite concentration changes. Typically, the metabolites identified from the S-plot are consistent with the major changes observed in the 2D ^{1}H–^{13}C HSQC spectrum. In fact, the 2D ^{1}H–^{13}C HSQC spectrum is used to aid the assignments of metabolites from the S-plots because of the large overlap in 1D ^{1}H NMR spectra. Nevertheless, the results from the S-plots and 2D ^{1}H–^{13}C HSQC spectrum need not be identical since the 2D ^{1}H–^{13}C HSQC experiments only monitor a subset of the metabolome defined by the ^{13}C-labeled metabolite used in the culture medium. Conversely, the 1D ^{1}H NMR spectra capture all the major metabolites (>10 μM) present in the metabolome.

2 Materials

2.1 Growth Medium and Cultivation Devices

1. Bacterial strain: *S. epidermidis* strain 1457 [41].
2. $^{13}C_6$-glucose (Cambridge Isotope Laboratories, Andover, MA, USA) or non-labeled glucose (0.25 % w/v) is added to bacto tryptic soy broth without dextrose (BD, Franklin Lakes, NJ, USA). The dry culture medium plus glucose is dissolved in ultrapure water (18.2 MΩ resistance) and filter sterilized using a 0.22 μm rapid-flow sterile disposable filter unit (*see* **Note 3**).
3. *S. epidermidis* are cultivated in Nalgene polypropylene Erlenmeyer flasks or in 14 mL disposable culture tubes.
4. Shaking incubator.
5. Spectrophotometer to monitor growth.

2.2 Reagents and Devices for Metabolomic Sample Preparation

1. 20 mM phosphate buffer, pH 7.2–7.4.

For 500 mL:

   (a) 1.045 g Dipotassium hydrogen phosphate.

   (b) 0.527 g Potassium dihydrogen phosphate.

   Dissolve salts in ultrapure water (18.2 MΩ resistance), and adjust volume to 500 mL.

2. Millipore Microfil V filtration devices (EMD Millipore, Billerica, MA, USA), filtration manifold (Thermo Fisher Scientific), vacuum pump (KNF Neuberger, Inc., Trenton, NJ, USA), and filter forceps.
3. Corning 50 mL graduated plastic centrifuge tubes (Corning, Corning, NY, USA).
4. Liquid nitrogen and ice.

5. FastPrep-24 instrument and FastPrep lysing matrix B tubes (MP Biomedicals, Solon, OH, USA).
6. Refrigerated microcentrifuge and 1.5 and 2.0 mL microcentrifuge tubes.
7. 20 mM phosphate buffer in 100 % D_2O (Sigma-Aldrich) at pH 7.2 (uncorrected).
8. Lyophilizer.
9. TMSP-d4.

2.3 NMR Instrumentation

1. Bruker 500-MHz Avance spectrometer equipped with a triple-resonance, z-axis gradient cryoprobe and an automatic tune and match (ATM) system.
2. BACS-120 sample changer with Bruker Icon software.

2.4 Data Analysis Software

1. SIMCA 11.5+ (UMETRICS) for PCA and OPLS-DA.
2. NMRViewJ [42] and Sparky (T. D. Goddard and D. G. Kneller, SPARKY 3, University of California, San Francisco) for NMR processing and analysis.
3. Metabominer, Madison Metabolomics Consortium Database (MMCD), the BioMagResBank (BMRB), and HMDB were used to identify metabolites [38–40, 43].
4. KEGG and Metacyc databases were used to verify metabolites and metabolic pathways [44, 45].
5. Cytoscape and Metscape were used to generate metabolic network maps [46, 47].

3 Methods

3.1 Bacterial Growth

1. Two days prior to the planned experiment, streak a TSB agar plate with *S. epidermidis* from a glycerol stock stored at −80 °C. Incubate overnight at 37 °C.
2. Inoculate a 14 mL disposable culture tube containing 2 mL of TSB with a single colony of *S. epidermidis* (*see* **Note 4**) and incubate at 37 °C on a 45° angle with 225 rpm of agitation. Cultivate overnight. This will be the culture used for starting the pre-culture.
3. Inoculate 25 mL of pre-warmed TSB, containing 0.25 % (w/v) glucose, in a 250 mL flask with 250 μl of the overnight culture from **step 2** in Subheading 3.1 and incubate at 37 °C with 225 rpm of agitation. Cultivate for 1.5–2 h. This will be the culture used for starting the primary cultures (*see* **Note 5**).
4. While the pre-culture is incubating, prepare the primary cultures for inoculation. Typically, we harvest 20 $O.D._{600}$ units for each growth condition, strain, or growth phase; therefore, the

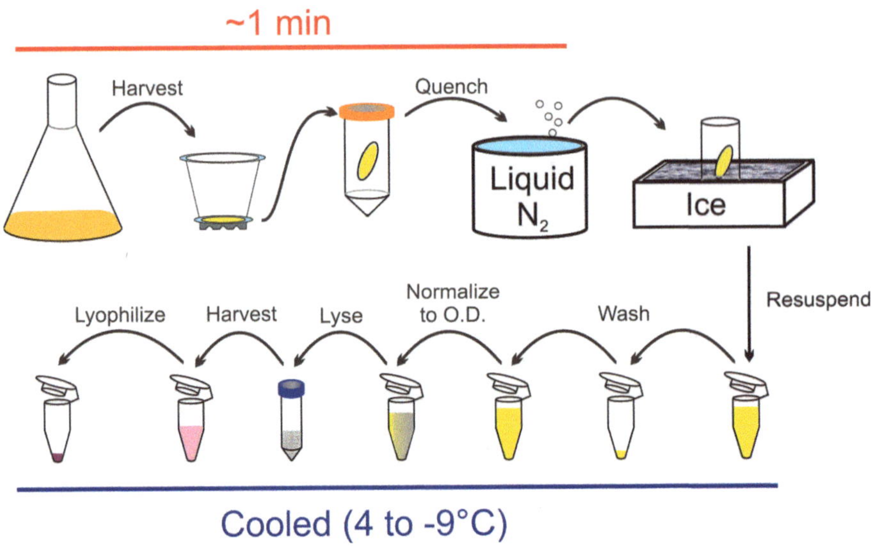

Fig. 7 A schematic representation of the sampling and preparation of *S. epidermidis* cell-free lysates for use in NMR metabolomics

volume of the primary culture will depend on the achievable optical density for the specific circumstances (*see* **Note 6**). Choose an appropriately sized flask to accommodate the volume of culture medium so as to maintain the correct flask-to-medium ratio. Add culture medium to the flasks, and pre-warm the medium to 37 °C (*see* **Note 7**).

5. After ~2 h of incubating the pre-culture (*see* **step 3** in Subheading 3.1), remove an aliquot and determine the $O.D._{600}$. Normalize the inocula so as to have a uniform starting $O.D._{600}$ of 0.06 in the primary cultures (*see* **Note 8**). Inoculate the pre-warmed primary cultures with the appropriate volume of pre-culture, swirl the flasks, remove a 1 mL sample and place these into individual microcentrifuge tubes, put primary cultures into the incubator, start agitation, and then verify that the starting optical densities were 0.06 ± 0.02 $O.D._{600}$ units.

3.2 Harvest (Summarized in Fig. 7)

1. Assemble the filtration apparatus, and pre-wash the filter with the non-deuterated phosphate buffer (*see* **item 1** in Subheading 2.2).
2. Harvest the volume of culture that is predicted to yield slightly (10 %) more than 20 $O.D._{600}$ units. At this stage, speed is more important than accuracy. Harvesting more than is required will compensate for bacterial loss during the wash steps. In addition, the optical density will be normalized in subsequent steps.
3. Disassemble the Millipore Microfil V filtration device, and, using filter forceps, transfer the filter to a Corning 50 mL graduated tube (precooled to −70 °C). Close the 50 mL tube and

immediately place it into a Dewar containing liquid nitrogen (*see* **Note 9**). Samples can remain in liquid nitrogen as other samples are being processed.

4. Transfer the Corning 50 mL graduated tubes to an ice bucket, and wash/scrape the bacteria off the filter using 1 mL of ice-cold phosphate buffer (*see* **item 1** in Subheading 2.2) by repeatedly pipetting the buffer over the filter.
5. Transfer the bacterial suspension to a 1.5 mL microcentrifuge tube (precooled to –20 °C), place the microcentrifuge tubes into a microcentrifuge precooled to –9 °C, and centrifuge for 1 min at 16,100 RCF.
6. After 1 min, place the microcentrifuge tubes into an ice bucket, and then remove and discard the supernatants. Suspend the cell pellet in 1 mL of ice-cold phosphate buffer and hold on ice (*see* **Note 10**).
7. Remove a 5 μl aliquot and add this to 995 μl of phosphate buffer (room temperature) in a disposable spectrophotometer cuvette, mix, and measure the $O.D._{600}$. Normalize the optical densities of the samples using ice-cold phosphate buffer such that all samples contain 20 $O.D._{600}$ units and have a 1 mL final volume.
8. Transfer the samples to FastPrep lysing matrix B tubes that have been cooled to –20 °C, and place the tubes in a FastPrep instrument. Lyse the bacteria at a speed of 6 for 40 s. Remove the tubes and place in an ice bucket to rest for 5 min, and then repeat the lysing step.
9. Centrifuge the lysing matrix B tubes for 2 min at 16,100 RCF in a microcentrifuge precooled to –9 °C.
10. Transfer 0.7 mL of the supernatant (cell-free lysate) to a pre-cooled (–20 °C) 2 mL microcentrifuge tube.
11. Add 1 mL of ice-cold phosphate buffer to each lysing matrix B tube, and suspend the pellet containing the glass beads and cell debris by repeated pipetting.
12. Centrifuge the lysing matrix B tubes for 2 min at 16,100 RCF in a microcentrifuge precooled to –9 °C.
13. Transfer 0.9 mL of the supernatant to the appropriate tubes from **step 10** in Subheading 3.2. The pooled volume of cell-free lysate should be 1.6 mL.
14. Centrifuge the pooled cell-free lysates for 1 min at 16,100 RCF in a microcentrifuge precooled to –9 °C.
15. Transfer 1.5 mL of the cell-free lysates to a new 2 mL microcentrifuge tubes that have been pre-chilled at –20 °C, and then freeze the samples in dry ice. This step will remove any glass beads and/or cell debris that were carried over from the lysing matrix B tubes.
16. Lyophilize the cell-free lysates at –50 °C.

3.3 Preparation and Data Acquisition of Metabolomic Samples by NMR

1. For 1D ^{1}H NMR, the lyophilized cell-free lysates (*see* **step 16** in Subheading 3.2) are suspended in 600 μl of 50 mM phosphate buffer (PBS) in 99.8 % D_2O at pH 7.2 (uncorrected), containing 50 μM TMSP-d4, which is transferred to an NMR tube. Typically, ten replicate samples are prepared for each class or cell condition to maximize statistical significance.
2. Our 1D ^{1}H NMR spectra are routinely collected on a Bruker 500 MHz Avance DRX spectrometer equipped with a cryoprobe to maximize sensitivity and using BACS-120 sample changer and auto-tune and match (ATM) for automation. 1D ^{1}H NMR spectra are collected at 298 K with a spectrum width of 5,482.5 Hz and 32 K data points. A total of 16 dummy scans and 64 scans are used to obtain each spectrum.
3. As with the 1D ^{1}H experiments, the 2D ^{1}H–^{13}C HSQC spectra are collected in an automated fashion on a Bruker 500 MHz Avance DRX spectrometer with a cryoprobe. The 2D ^{1}H–^{13}C HSQC spectra are collected with solvent presaturation and a relaxation delay of 0.5 s. A total of 2,048 data points with a spectrum width of 4,734.9 Hz and 128 data points with a spectrum width of 13,834.3 Hz are collected in the ^{1}H and ^{13}C dimensions, respectively. A total of 8 dummy scans and 128 scans are used to obtain each of the 2D ^{1}H–^{13}C HSQC NMR spectra.

3.4 NMR Metabolomic Data Analysis

Due to the wide array of methods, protocols, and software available for NMR data analysis, we have not included a detailed protocol describing our approach to data analysis. Instead, we have included in the introductory section some general considerations to keep in mind when doing NMR data analysis (*see* Subheading 1.8).

4 Notes

1. When assessing bacterial numbers, it is important to remember that a certain percentage of bacteria will be nonviable or dead, leading to an overestimation in the number of live bacteria. Usually, this does not create a significant obstacle unless the bacteria are being harvested in the stationary or the death phases or when using mutants that affect cell growth, division, and/or autolysis. Under circumstances where viability may be compromised, the percentage of live bacteria must be determined using methods such as live–dead stains.
2. PBS made with D_2O is used as the solvent because there are no observable hydrogen nuclei in the buffer due to the fact that the majority of hydrogen atoms have been replaced by deuterium.
3. Filter sterilizing culture media prevents pyrolysis of heat labile amino acids and carbohydrates, such as that which occurs

during autoclaving. Pyrolysis is particularly problematic when differences are introduced into culture media autoclaving procedures, specifically, differences in heat exposure time and load density. For these reasons, our experience is that filter sterilization increases the consistency of growth profiles and produces a more uniform metabolome sample.

4. As staphylococci are prone to acquiring mutations during passage and storage [48], use colonies from freshly streaked plates. Using colonies from plates that have been stored at room temperature or 4 °C for more than a few days increases the risk that adaptive mutagenesis will have induced undesirable mutations.
5. By using a pre-culture that is in the exponential growth phase (i.e., has been growing for only 1.5–2 h), the lag phase will be minimal because the bacteria reached their initiation mass during the pre-culture.
6. Prior to initiating metabolomic studies, growth rates and optical densities must be determined empirically for each growth condition, strain, or growth phase. With this information it is possible to estimate the volume of culture that will need to be harvested; however, the actual number of $O.D._{600}$ units used in preparing the metabolome samples will be adjusted based on the optical density. As an example, if we want to examine the metabolome of *S. epidermidis* strain 1457 during the exponential growth phase and we have determined that strain 1457 can achieve an $O.D._{600}$ of 0.4 (Fig. 2), then it is necessary to harvest 50 mL to reach 20 $O.D._{600}$ units.
7. Pre-warming the culture medium will prevent a temperature shock during inoculation.
8. To prevent the introduction of "too much" non-labeled glucose into the primary culture medium, it may be preferable to concentrate the bacteria by centrifugation and suspend in a smaller volume of pre-warmed medium.
9. As discussed, speed is critical for achieving an accurate representation of the metabolome; hence, **steps 2** and **3** in Subheading 3.2 combined should take about 1 min.
10. Washing the bacteria is essential for the removal of metabolites found in the culture medium, which can introduce artifacts into the NMR spectra.

Acknowledgements

G.A.S. and R.P. were supported by funds provided through the National Institutes of Health (AI087668). R.P. was also supported by funds through the National Institutes of Health National Center for Research Resources (P20 RR-17675) and the American Heart

Association (0860033Z). The research was performed in facilities renovated with support from the National Institutes of Health (NIH, RR015468-01). We are grateful to our past and present students and postdoctoral fellows for their hard work and assistance—thanks, for all that you do!

References

1. Neidhardt FC (2006) Apples, oranges and unknown fruit. Nat Rev Microbiol 4:876
2. Somerville GA, Proctor RA (2009) At the crossroads of bacterial metabolism and virulence factor synthesis in Staphylococci. Microbiol Mol Biol Rev 73:233–248
3. Neidhardt FC, Bloch PL, Smith DF (1974) Culture medium for enterobacteria. J Bacteriol 119:736–747
4. Hussain M, Hastings JG, White PJ (1991) A chemically defined medium for slime production by coagulase-negative staphylococci. J Med Microbiol 34:143–147
5. Novick RP (1991) Genetic systems in staphylococci. Methods Enzymol 204:587–636
6. Sadykov MR, Olson ME, Halouska S et al (2008) Tricarboxylic acid cycle-dependent regulation of *Staphylococcus epidermidis* polysaccharide intercellular adhesin synthesis. J Bacteriol 190:7621–7632
7. Sadykov MR, Zhang B, Halouska S et al (2010) Using NMR metabolomics to investigate tricarboxylic acid cycle dependent signal transduction in *Staphylococcus epidermidis*. J Biol Chem 285:36616–36624
8. Somerville GA, Said-Salim B, Wickman JM et al (2003) Correlation of acetate catabolism and growth yield in *Staphylococcus aureus*: implications for host-pathogen interactions. Infect Immun 71:4724–4732
9. Sadykov MR, Hartmann T, Mattes TA et al (2011) CcpA coordinates central metabolism and biofilm formation in *Staphylococcus epidermidis*. Microbiology 157:3458–3468
10. Krebs HA (1972) The Pasteur effect and the relations between respiration and fermentation. Essays Biochem 8:1–34
11. Miles AA, Misra SS, Irwin JO (1938) The estimation of the bactericidal power of the blood. J Hyg (Lond) 38:732–749
12. van Gulik WM (2010) Fast sampling for quantitative microbial metabolomics. Curr Opin Biotechnol 21:27–34
13. Zhang B, Powers R (2012) Analysis of bacterial biofilms using NMR-based metabolomics. Future Med Chem 4:1273–1306
14. Hwang T-L, Shaka AJ (1995) Water suppression that works. Excitation sculpting using arbitrary waveforms and pulsed field gradients. J Magn Reson A 112:275–279
15. Beckonert O, Keun HC, Ebbels TM et al (2007) Metabolic profiling, metabolomic and metabonomic procedures for NMR spectroscopy of urine, plasma, serum and tissue extracts. Nat Protoc 2:2692–2703
16. Meiboom S, Gill D (1958) Modified spin-echo method for measuring nuclear relaxation times. Rev Sci Intrum 29:4
17. Bingol K, Bruschweiler R (2011) Deconvolution of chemical mixtures with high complexity by NMR consensus trace clustering. Anal Chem 83:7412–7417
18. Sands CJ, Coen M, Ebbels TM et al (2011) Data-driven approach for metabolite relationship recovery in biological 1H NMR data sets using iterative statistical total correlation spectroscopy. Anal Chem 83:2075–2082
19. Craig A, Cloarec O, Holmes E et al (2006) Scaling and normalization effects in NMR spectroscopic metabonomic data sets. Anal Chem 78:2262–2267
20. Halouska S, Powers R (2006) Negative impact of noise on the principal component analysis of NMR data. J Magn Reson 178: 88–95
21. Sysi-Aho M, Katajamaa M, Yetukuri L et al (2007) Normalization method for metabolomics data using optimal selection of multiple internal standards. BMC Bioinformatics 8:93
22. van den Berg RA, Hoefsloot HC, Westerhuis JA et al (2006) Centering, scaling, and transformations: improving the biological information content of metabolomics data. BMC Genomics 7:142
23. Anderson PE et al (2011) Dynamic adaptive binning: an improved quantification technique for NMR spectroscopic data. Metabolomics 7:179–190
24. Anderson PE et al (2008) Gaussian binning: a new kernel-based method for processing NMR spectroscopic data for metabolomics. Metabolomics 4:261–272

25. Davis RA et al (2007) Adaptive binning: an improved binning method for metabolomics data using the undecimated wavelet transform. Chemometr Intell Lab Syst 85:144–154
26. De Meyer T, Sinnaeve D, Van Gasse B et al (2008) NMR-based characterization of metabolic alterations in hypertension using an adaptive, intelligent binning algorithm. Anal Chem 80:3783–3790
27. Bylesjo M et al (2006) OPLS discriminant analysis: combining the strengths of PLS-DA and SIMCA classification. J Chemometr 20: 341–351
28. Barker M, Rayens W (2003) Partial least squares for discrimination. J Chemometr 17: 166–173
29. Rannar S et al (1994) A Pls Kernel algorithm for data sets with many variables and fewer objects. 1. Theory and algorithm. J Chemometr 8:111–125
30. Werth MT, Halouska S, Shortridge MD et al (2010) Analysis of metabolomic PCA data using tree diagrams. Anal Biochem 399:58–63
31. Worley B, Halouska S, Powers R (2013) Utilities for quantifying separation in PCA/PLS-DA scores plots. Anal Biochem 15:102–104
32. Kjeldahl K, Bro R (2010) Some common misunderstandings in chemometrics. J Chemometr 24:558–564
33. Golbraikh A, Tropsha A (2002) Beware of q(2)! J Mol Graph Model 20:269–276
34. Shao J (1993) Linear-model selection by cross-validation. J Am Stat Assoc 88:486–494
35. Westerhuis JA et al (2008) Assessment of PLSDA cross validation. Metabolomics 4:81–89
36. Eriksson L, Trygg J, Wold S (2008) CV-ANOVA for significance testing of PLS and OPLS (R) models. J Chemometr 22:594–600
37. Hu K, Westler WM, Markley JL (2011) Simultaneous quantification and identification of individual chemicals in metabolite mixtures by two-dimensional extrapolated time-zero (1) H-(13)C HSQC (HSQC(0)). J Am Chem Soc 133:1662–1665
38. Cui Q, Lewis IA, Hegeman AD et al (2008) Metabolite identification via the Madison Metabolomics Consortium Database. Nat Biotechnol 26:162–164
39. Ulrich EL, Akutsu H, Doreleijers JF et al (2008) BioMagResBank. Nucleic Acids Res 36:D402–D408
40. Wishart DS, Jewison T, Guo AC et al (2007) HMDB: the Human Metabolome Database. Nucleic Acids Res 35:D521–D526
41. Mack D, Siemssen N, Laufs R (1992) Parallel induction by glucose of adherence and a polysaccharide antigen specific for plastic-adherent *Staphylococcus epidermidis*: evidence for functional relation to intercellular adhesion. Infect Immun 60:2048–2057
42. Johnson BA (2004) Using NMRView to visualize and analyze the NMR spectra of macromolecules. Methods Mol Biol 278:313–352
43. Xia J, Bjorndahl TC, Tang P et al (2008) MetaboMiner—semi-automated identification of metabolites from 2D NMR spectra of complex biofluids. BMC Bioinformatics 9:507
44. Caspi R, Altman T, Dale JM et al (2010) The MetaCyc database of metabolic pathways and enzymes and the BioCyc collection of pathway/genome databases. Nucleic Acids Res 38: D473–D479
45. Kanehisa M, Araki M, Goto S et al (2008) KEGG for linking genomes to life and the environment. Nucleic Acids Res 36:D480–D484
46. Karnovsky A, Weymouth T, Hull T et al (2012) Metscape 2 bioinformatics tool for the analysis and visualization of metabolomics and gene expression data. Bioinformatics 28:373–380
47. Killcoyne S, Carter GW, Smith J et al (2009) Cytoscape: a community-based framework for network modeling. Methods Mol Biol 563: 219–239
48. Somerville GA, Beres SB, Fitzgerald JR et al (2002) In vitro serial passage of *Staphylococcus aureus*: changes in physiology, virulence factor production, and *agr* nucleotide sequence. J Bacteriol 184:1430–1437
49. Zhang B, Halouska S, Schiaffo CE et al (2011) NMR analysis of a stress response metabolic signaling network. J Proteome Res 10: 3743–3754

Chapter 7

The Isolation and Analysis of Phenol-Soluble Modulins of *Staphylococcus epidermidis*

Hwang-Soo Joo and Michael Otto

Abstract

Phenol-soluble modulins (PSMs) are multifunctional peptide toxins produced by many staphylococcal strains. PSMs have received much recent attention, owing to multiple reports underscoring their importance for staphylococcal pathogenesis. Members of the PSM family may be strongly cytolytic to neutrophils and other cell types; promote inflammatory, receptor-mediated responses in several human cell types; and contribute to biofilm structuring and detachment. Here we describe biochemical methods to isolate, purify, and quantitatively analyze *Staphylococcus epidermidis* PSMs.

Key words *Staphylococcus epidermidis*, Phenol-soluble modulins, Peptide toxin, Liquid chromatography, Mass spectrometry

1 Introduction

Phenol-soluble modulins (PSMs) are multifunctional, amphipathic, α-helical peptides produced by virtually all staphylococcal strains. They are well known as major inflammatory and cytolytic toxins of *Staphylococcus epidermidis* and *S. aureus* [1–4]. In addition, they have antibacterial activity, likely to compete with other environmental rival bacteria such as streptococci [5–7], as well as biofilm structuring and dissemination functions [8, 9]. Interestingly, they also appear to protect biofilms from disassembly by forming extracellular amyloid-like fibril structures, at least under certain in vitro growth conditions [10].

S. epidermidis PSMs were first discovered in the phenol layer after hot aqueous phenol extraction of culture supernatant during the investigation of factors with pro-inflammatory activity toward a macrophage cell line [1]. Since that first report, which described three *S. epidermidis* PSMs (PSMα, PSMβ, PSMγ or δ-toxin), a total of six PSMs produced from the *S. epidermidis* genome (PSMα, PSMβ1, PSMβ2, PSMγ/δ-toxin, PSMδ, and PSMε) were found [1, 11, 12]. In addition, PSM-mec is encoded on some SCC*mec*

Paul D. Fey (ed.), *Staphylococcus Epidermidis: Methods and Protocols*, Methods in Molecular Biology, vol. 1106,
DOI 10.1007/978-1-62703-736-5_7, © Springer Science+Business Media, LLC 2014

methicillin resistance mobile genetic elements present in *S. aureus* and *S. epidermidis* [13, 14]. PSMs are virtually the only factors with toxic potential that *S. epidermidis* produces; therefore, they are believed to play an important role in *S. epidermidis* pathogenesis [15]. The PSMβ peptides were shown to promote biofilm structuring, detachment, and in vivo dissemination of biofilm-associated infection [8], while direct evidence for the role of other *S. epidermidis* PSMs using isogenic gene deletion mutants is still missing. However, some *S. epidermidis* PSMs, in particular PSMδ, are known to have strong cytolytic potential, as based on the analysis of pure peptides [3]. For a more detailed future investigation of the role of this important toxin family in *S. epidermidis* pathogenesis, it is necessary to possess a variety of analytical tools for PSMs. Here we describe a range of techniques for PSM isolation and analysis in *S. epidermidis* research, also to be used in similar form for other staphylococci.

2 Materials

1. Bact tryptic soy broth (TSB), supplemented with 0.5 % glucose after autoclave sterilization.
2. Ultrafiltration devices: Centricon Plus-70 (MWCO 10 kD) and Centriprep YM-10 (MWCO 10 kD) (Millipore, Bedford, MA).
3. Dialysis tubings: Spectra/Por CE (MWCO 20 kD) (Spectrum Laboratories, Rancho Dominguez, CA), dialysis tubing (MWCO 12 kD) (Sigma, St. Louis, MO).
4. Chromatography instruments: High-performance liquid chromatography/mass spectrometry (HPLC/MS) system (1100 HPLC system connected to an MSD Trap SL mass spectrometer, Agilent Technologies, Santa Clara, CA), AKTA Purifier system (GE Healthcare, Piscataway, NJ).
5. Chromatography columns: Resource PHE material column (GE Healthcare, Piscataway, NJ), Zorbax SB-C3 column (9.4 × 250 mm) and Zorbax SB-C8 column (2.3 × 30 mm) (Agilent Technologies, Santa Clara, CA), Vydac C4 column (4.6 × 150 mm) (Grace, Deerfield, IL).
6. Eluent A and B for one-step method for hot phenol-extracted crude PSMs: eluent A, 2.5 mM ammonium acetate, pH 6.7; eluent B, 100 % 1-propanol.
7. Eluent A and B for two-step method for TCA-precipitated crude PSMs: eluent A, 0.1 % TFA in 10 % acetonitrile/90 % water; eluent B, 0.1 % TFA in 90 % acetonitrile/10 % water.
8. Eluent A and B for quantitative analysis of PSMs from culture supernatants by LC/MS: eluent A, 0.1 % TFA in 100 % water; eluent B, 0.1 % TFA in acetonitrile.

9. Phosphate-buffered saline (PBS): 10 mM sodium phosphate, 150 mM NaCl, pH 7.4.
10. Tris buffer: 100 mM Tris–HCl, pH 8.0.
11. 1 M sodium acetate, pH 4.7.
12. 0.4 M sodium chloride.
13. 100 mM Na_2HPO_4.
14. Buffer-saturated phenol.
15. 100 % (w/v) trichloroacetic acid (TCA) solution.
16. Trifluoroacetic acid (TFA) acetonitrile.
17. 1-propanol.
18. 1-butanol.
19. Spectrophotometer.
20. Vacuum centrifuge.
21. Lyophilizer.
22. *Staphylococcus epidermidis* strains.
23. QuantAnalysis software (Agilent): For use with LC/MS.

3 Methods

Pre-cultures grown overnight are used to inoculate bacterial cultures at 1:100 dilution or an initial OD_{600nm} of 0.1, unless noted otherwise.

3.1 Isolation of Crude PSMs

Crude PSMs isolated by the methods described here can be applied for further purification of PSMs or directly used for biochemical and immunological research. However, it must be noted that these purification procedures do not allow complete separation from other pro-inflammatory products. For the analysis of pro-inflammatory activities, pure (synthetic) PSM peptides should be used. There are three described methods to isolate crude PSMs, which use hot phenol, TCA, or butanol [1, 7, 16]. All steps are conducted at 4 °C unless otherwise noted.

3.1.1 Hot Phenol Extraction Method

1. Remove *S. epidermidis* cells by centrifugation (6,000 × *g*, 30 min, 4 °C) after overnight growth (~16 h) in 100 mL of TSB or appropriate media at 37 °C with shaking at 180 rpm.
2. Clarify supernatants by filtration.
3. Concentrate the filtrate further by centrifugal ultrafiltration (Centricon Plus-70, MWCO 10 kD) to ~1 mL (*see* **Note 1**).
4. Dialyze the concentrate against 0.4 M NaCl in MWCO 20 kD tubing.

5. Dialyze the filtrate further against distilled water in MWCO 12 kD dialysis tubing.
6. Add 2.5 mL of buffer-saturated phenol and enough volume of 1 M sodium acetate (pH 4.7) to the retentate to make the aqueous portion to 0.1 M.
7. Agitate the mixture at 65 °C for 1 h maintaining a single phase.
8. Cool down the solution and centrifuge for 15 min.
9. Collect the phenol layer, and extract the aqueous layer twice more with 1.25 mL phenol.
10. Dialyze the pooled phenol layer in distilled water at 4 °C in 12-kD MWCO tubing.
11. Vigorously mix the fine precipitate (PSMs) into the aqueous retentate.
12. Lyophilize the solution and store at −20 °C.

*3.1.2 TCA Precipitation Method (See **Note 2**)*

1. Prepare culture supernatants as described in **steps 1** and **2** in Subheading 3.1.1.
2. Add 1/9 (v/v) of ice-cold 100 % TCA into the supernatants to reach 10 % (v/v) of TCA.
3. Incubate the solutions at 4 °C for 3 h.
4. Harvest the precipitates by centrifugation (6,000 × *g*, 20 min, 4 °C).
5. Dissolve the precipitates in 100 mM Na_2HPO_4 and adjust pH between 7.0 and 7.5 by adding 10 *N* NaOH.
6. After brief centrifugation, concentrate the supernatants five-fold by centrifugal ultrafiltration (Centriprep YM-10).
7. Lyophilize the retentate and store at −20 °C.

3.1.3 Butanol Extraction Method

1. Prepare culture supernatants as described in **steps 1** and **2** in Subheading 3.1.1.
2. Add 1/3 (v/v) of 100 % 1-butanol into the supernatants to make 25 % (v/v) of 1-butanol.
3. Shake the solutions vigorously at 37 °C for 2 h.
4. After brief centrifugation, collect upper (butanol) phases.
5. Dry the solutions by vacuum centrifugation and store at −20 °C.

3.2 Purification of PSMs by Liquid Chromatography

Due to their amphipathic α-helical structure, PSM peptides show surfactant-like property. This causes longer retention times on reversed-phase columns during liquid chromatography compared to most other proteins in the culture supernatant. This makes it relatively simple to purify and analyze PSMs by reversed-phase chromatography.

Several methods using different LC systems and solvents were reported for column-based PSM purification, usually relying on one or more reversed-phase chromatography steps. Any similar protocol can be used according to the complexity of crude PSM isolates.

3.2.1 One-Step Method for Hot Phenol-Extracted Crude PSMs (Modified from Ref. 1)

1. Dissolve 100 μg of lyophilized crude PSM extract in 20 μL of 20 % 1-propanol and inject it into AKTA Purifier system with Vydac C4 column.
2. Use the following method with a UV detector at 214 nm to separate and fractionate PSMs:
   (a) Flow rate, 1 mL/min; eluent A, 2.5 mM ammonium acetate, pH 6.7; eluent B, 100 % 1-propanol.
   (b) 28–48 % linear gradient of eluent B for 1 h.
3. The fractions are collected and lyophilized. The lyophilized fraction is dissolved in PBS by vigorous vortexing for subsequent assays.
4. PSM fractions can normally be detected between 35 and 40 % of eluent B gradient by using an HIV-1 LTR luciferase assay [17] or mass spectrometry.

3.2.2 Two-Step Method for TCA-Precipitated Crude PSMs (Modified from Refs. 12, 16)

1. Dissolve the lyophilized crude PSMs or the TCA precipitate in 100 mM Tris buffer (pH 8.0) and adjust pH to neutral with 10 *N* NaOH.
2. Use self-packed Resource PHE material column (GE Healthcare, Piscataway, NJ) with the following HPLC method.
3. Flow rate, 4 mL/min; eluent A, 0.1 % TFA in 10 % acetonitrile/90 % water; eluent B, 0.1 % TFA in 90 % acetonitrile/10 % water.
4. 3 column volumes (cv) of 100 % eluent A.
5. 3 cv of 50 % eluent A:50 % eluent B.
6. 10 cv of gradient with 50–100 % eluent B.
7. Collect and concentrate the PSM fractions and inject into a Zorbax SB-C3 column.
8. Use the same HPLC method as in **step 2**.

3.3 Quantitative Analysis of PSMs from Culture Supernatants by LC/MS (Modified from Refs. 2, 16)

While PSMs are well separated from most other proteins in the culture filtrate, detailed quantitative analysis using UV detection is difficult, because different PSMs do not completely separate during reversed-phase chromatography. For that reason, HPLC/MS procedures are commonly used to quantify PSM production. Here we describe the reversed-phase HPLC/ESI-MS method used in our laboratory; other reversed-phase columns and mass spectrometry methods (such as MALDI-TOF) may also be used.

1. Grow staphylococcal cells in TSB or appropriate media to reach the purposed growth phase.
2. Take 1 mL of culture into microcentrifuge tubes.
3. Centrifuge in a cooled tabletop microcentrifuge at maximal speed for 10 min, and transfer the supernatants into glass sample vials (*see* **Note 3**).
4. Cool down the sample tray to 10 °C and put the sample vials in it.
5. Appropriate volumes of samples between 20 and 100 μL may be injected into a Zorbax SB-C8 2.3 × 30 mm reversed-phase column using an Agilent 1100 HPLC system coupled to an Agilent Trap SL ion trap mass spectrometer.
6. The following gradient program can be used for the rapid detection of PSMs:
   (a) Flow rate, 0.5 mL/min; eluent A, 0.1 % TFA in 100 % water; eluent B, 0.1 % TFA in acetonitrile.
   (b) 10 % eluent B for 2.5 min; 50 % eluent B for 2.5 min; a linear gradient from 50 to 90 % of eluent B for 10 min.
7. PSMs can generally be detected between 55 and 75 % of eluent B. *S. epidermidis* PSMs normally elute at 60–65 % of eluent B.
8. Depending on the mass spectrometer settings, the amount of each PSM is determined by integration of the extracted ion chromatogram of the main two peaks of each PSM (using our settings; doubly and triply charged ion peaks for PSMα, PSMγ, PSMδ, and PSMε; triply and quadruply charged ion peaks for PSMβ).
9. Absolute amounts of PSMs can be determined by using calibration curves made from each synthetic PSM. Semiautomatic quantification can be done by using QuantAnalysis software with a preset of calculated m/z ratios and estimated retention time of each PSM (*see* **Note 4**).

4 Notes

1. Ultrafiltration using tangential flow filter (TFF) cartridge NMWL (10 kD) can be applied before centrifugal ultrafiltration for larger culture volumes. Although the molecular weights of PSMs are between 2 and 5 kD, larger MWCO membranes may be used with care, because PSMs form a complex aggregate of about ~35 kD under certain conditions [1, 16].
2. Methionine sulfoxidation may happen to a small amount of PSMs during TCA precipitation, which can produce additional derivatives having 16 Da higher molecular weight [16].
3. Some staphylococcal cells are difficult to pellet under certain conditions. Speed and time of centrifugation may vary according

Table 1
***m/z* ratios and retention times of *S. epidermidis* PSMs**

PSM	*m/z*	RT (min)/eluent B (%)
PSMα	1,245.0 (d), 830.3 (t)	7.6/60.4
PSMβ1	1,556.8 (t), 1167.8 (q)	7.6/60.4
PSMγ	1,425.2 (d), 950.5 (t)	7.8/61.4
PSMδ	1,324.6 (d), 883.4 (t)	7.5/60
PSMε	1,399.2 (d), 933.1 (t)	8.4/63.6

to the aggregating properties of the cells. Culture supernatant samples in glass sample vials can be stored at 4 °C for a day or in −20 °C for longer periods. However, repeated freezing and thawing cycles can precipitate PSMs or possibly produce vial-attached polymeric PSM forms, which results in decreased amounts of LC/MS-detectable PSMs. We found that using strongly denaturing buffers containing 8 M urea can be used to overcome those problems.

4. Table 1 shows an example of the preset of *m/z* ratios and retention times of *S. epidermidis* PSMs for quantification. Retention times may vary according to the LC system and the tubing length to MS detector. (d), (t), and (q) stand for doubly, triply, and quadruply charged ions, respectively.

Acknowledgements

This work was supported by the Intramural Research Program of the National Institute of Allergy and Infectious Diseases, National Institutes of Health.

References

1. Mehlin C, Headley CM, Klebanoff SJ (1999) An inflammatory polypeptide complex from *Staphylococcus epidermidis*: isolation and characterization. J Exp Med 189:907–918
2. Wang R, Braughton KR, Kretschmer D et al (2007) Identification of novel cytolytic peptides as key virulence determinants for community-associated MRSA. Nat Med 13: 1510–1514
3. Cheung GY, Duong AC, Otto M (2012) Direct and synergistic hemolysis caused by *Staphylococcus* phenol-soluble modulins: implications for diagnosis and pathogenesis. Microbes Infect 14:380–386
4. Cheung GY, Rigby K, Wang R et al (2010) *Staphylococcus epidermidis* strategies to avoid killing by human neutrophils. PLoS Pathog 6:e1001133
5. Cogen AL, Yamasaki K, Muto J et al (2010) *Staphylococcus epidermidis* antimicrobial delta-toxin (phenol-soluble modulin-gamma) cooperates with host antimicrobial peptides to kill group A *Streptococcus*. PLoS ONE 5:e8557
6. Cogen AL, Yamasaki K, Sanchez KM et al (2010) Selective antimicrobial action is provided by phenol-soluble modulins derived from *Staphylococcus epidermidis*, a normal resident of the skin. J Invest Dermatol 130:192–200

7. Joo HS, Cheung GY, Otto M (2011) Antimicrobial activity of community-associated methicillin-resistant *Staphylococcus aureus* is caused by phenol-soluble modulin derivatives. J Biol Chem 286:8933–8940
8. Wang R, Khan BA, Cheung GY et al (2011) *Staphylococcus epidermidis* surfactant peptides promote biofilm maturation and dissemination of biofilm-associated infection in mice. J Clin Invest 121:238–248
9. Periasamy S, Joo HS, Duong AC et al (2012) How *Staphylococcus aureus* biofilms develop their characteristic structure. Proc Natl Acad Sci U S A 109:1281–1286
10. Schwartz K, Syed AK, Stephenson RE et al (2012) Functional amyloids composed of phenol soluble modulins stabilize *Staphylococcus aureus* biofilms. PLoS Pathog 8:e1002744
11. Vuong C, Durr M, Carmody AB et al (2004) Regulated expression of pathogen-associated molecular pattern molecules in *Staphylococcus epidermidis*: quorum-sensing determines proinflammatory capacity and production of phenol-soluble modulins. Cell Microbiol 6: 753–759
12. Yao Y, Sturdevant DE, Otto M (2005) Genomewide analysis of gene expression in *Staphylococcus epidermidis* biofilms: insights into the pathophysiology of *S. epidermidis* biofilms and the role of phenol-soluble modulins in formation of biofilms. J Infect Dis 191: 289–298
13. Queck SY, Khan BA, Wang R et al (2009) Mobile genetic element-encoded cytolysin connects virulence to methicillin resistance in MRSA. PLoS Pathog 5:e1000533
14. Chatterjee SS, Chen L, Joo HS et al (2011) Distribution and regulation of the mobile genetic element-encoded phenol-soluble modulin PSM-mec in methicillin-resistant *Staphylococcus aureus*. PLoS ONE 6:e28781
15. Otto M (2009) *Staphylococcus epidermidis*—the 'accidental' pathogen. Nat Rev Microbiol 7:555–567
16. Otto M, O'Mahoney DS, Guina T et al (2004) Activity of *Staphylococcus epidermidis* phenol-soluble modulin peptides expressed in *Staphylococcus carnosus*. J Infect Dis 190: 748–755
17. Klebanoff SJ, Kazazi F, Van Voorhis WC et al (1994) Activation of the human immunodeficiency virus long terminal repeat in THP-1 cells by a staphylococcal extracellular product. Proc Natl Acad Sci U S A 91:10615–10619

Chapter 8

Genetic Manipulation of Staphylococci

Jeffrey L. Bose

Abstract

The ability to genetically manipulate bacteria is essential to understanding gene/protein function in these organisms. While basic cloning has become routine in molecular biology, many still view the ability to make directed mutations as a daunting or intimidating task. To aid the staphylococcal research community, the goal of this treatise is to describe the method of allelic exchange using temperature-sensitive plasmids that we have used to successfully produce a variety of mutations including single nucleotide changes in the *Staphylococcus aureus* chromosome. In addition, this chapter provides extensive summaries to aid in the construction of mutations, complementation plasmids, as well as transcription and translation reporters.

Key words Allelic exchange, Mutant, Mutation, Reporter, Complement

1 Introduction

From early studies of UV or chemical mutagenesis to the modern-day ability to generate targeted single nucleotide changes, the ability to mutate bacteria has proven essential to understanding the physiology of these organisms. While the tools described in this chapter have been demonstrated in *Staphylococcus aureus*, most of these methods can be directly applied to *Staphylococcus epidermidis*. Individual sections will have their own introductions covering necessary background information and specific adjustments needed for *S. epidermidis* are highlighted in Subheading 4. Allelic exchange is the act of replacing a piece of DNA (allele) with an alternate DNA fragment such as a mutation (mutant allele) via homologous recombination. Several factors are important for allelic exchange to occur. First, the cell must have an active recombination system. Second, recombination occurs more frequently when there is more homologous DNA (*see* **Note 1**). Allelic exchange plasmids for Staphylococci are shuttle vectors that contain *Escherichia coli* elements to facilitate cloning and *S. aureus* factors to aid in selection and/or screening. These plasmids contain a temperature-sensitive origin of replication derived from pE194 such that they replicate at

Paul D. Fey (ed.), *Staphylococcus Epidermidis: Methods and Protocols*, Methods in Molecular Biology, vol. 1106,
DOI 10.1007/978-1-62703-736-5_8, © Springer Science+Business Media, LLC 2014

the permissive temperature of 30 °C but not at higher temperatures. Therefore, a temperature shift is used to control plasmid replication and integration. In addition, they must contain the desired mutation to be placed on the chromosome. We generally consider three types of mutations, including (1) point mutations, (2) unmarked deletions, and (3) deletions with an insertion of foreign DNA. In order to make targeted mutations, one needs to understand the concepts of making mutations and the process of allelic exchange via recombination. In addition, to fully characterize generated mutants, other methods such as complementation and reporters of gene expression are often necessary. Therefore, this chapter contains an extensive Notes section covering these topics.

2 Materials

1. Tryptic Soy Agar (TSA).
2. Tryptic Soy Broth (TSB).
3. Antibiotics as needed.
4. Appropriate plasmids. *See* Table 1 for suggested plasmids.
5. Polymerases for PCR. KOD (Novagen, Madison, WI) is suggested.
6. 37 and 30 °C shaking incubator.
7. 44 °C non-shaking incubator.

Table 1
Suggested plasmids

	Characteristics[a]			
Plasmid	***Staphylococcus***	***E. coli***	**Uses**	**Source**
pCL10	oriV^{TS}, Chl^r	Amp^r oriV^{HC}	Allelic exchange	[12]
pCL52.2	oriV^{TS}, Tet^r	Spt^r oriV^{LC}	Allelic exchange	[12]
pCN51	oriV, Ery^r, P_{cad}	Amp^r oriV^{HC}	Complementation/expression	[1]
pJB38	oriV^{TS}, Chl^r, Atet	Amp^r oriV^{HC}	Allelic exchange	[7]
pKOR1	oriV^{TS}, Chl^r, Atet	Chl^r, Amp^r, oriV^{HC}	Allelic exchange	[13]
pLI50	oriV, Chl^r	Amp^r, oriV^{LC}	Complementation	[14]
pRN8298	oriV^{LC}, Ery^r	Amp^r oriV^{HC}	Complementation	[1]

[a]Abbreviations are for temperature-sensitive origin (oriV^{TS}), low-copy origin (oriV^{LC}), high-copy origin (oriV^{HC}), cadmium-inducible promoter (P_{cad}), anhydrotetracycline-inducible counter-selection (Atet), and resistance to ampicillin (Amp^r), chloramphenicol (Chl^r), spectinomycin (Spt^r), and tetracycline (Tet^r)

3 Methods

3.1 Preparing Strain

1. Generate a mutagenesis plasmid using a temperature-sensitive plasmid containing the desired deletion, marked mutation, or point mutation (*see* **Note 2**).
2. Move the plasmid into a strain of interest, maintaining culture at 30 °C.

3.2 Allelic Exchange

Performing allelic exchange starts with moving the mutagenesis plasmid into the desired strain. The strains go through two temperature shifts to encourage recombination. The first is at high temperature (~44 °C) at which the plasmid does not replicate well and therefore colonies that are able to grow on selective media have undergone integration of the plasmid into the chromosome (Fig. 1, steps 1 and 2). The resulting cells are often called single recombinants (SRs). This step is followed by nonselective growth at the permissive temperature of 30 °C to encourage a second recombination event with the plasmid leaving the chromosome and then being lost as the cells replicate (Fig. 1, **step 3**).

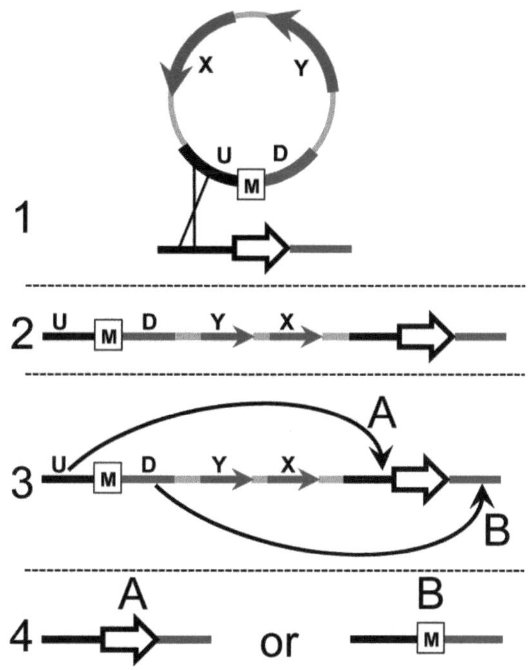

Fig. 1 Allelic exchange. First, homologous DNA in the upstream (U, *black*, shown) or downstream (D, *red*, not shown) of target gene and cloned sequence in plasmid undergo the first recombination event (1) leading to integration of plasmid in chromosome (2) and is antibiotic resistant. During the second event (3), recombination occurs between either the upstream (A) or downstream fragments (B). This results in restoration of wild-type (A) or mutant (B) genotype and is antibiotic sensitive (4)

These cells are called double recombinants and depending on where the recombination occurs, can lead to either a wild-type or mutant genotype (Fig. 1, step 4).

1. Day 1. In late afternoon, streak from −80 °C stock onto several TSA plates with antibiotics and place at 44 °C.
2. Day 2. In the morning, examine plates for large and small colonies. Re-streak large colonies because these should be cells with a single integration of the plasmid in the chromosome. *See* **Note 3** for *S. epidermidis.*
3. Day 3. Examine plates and make sure that they grew well to confirm that they are SRs. Inoculate several colonies individually into 3 ml of TSB and incubate at 30 °C overnight. *See* **Note 4** for preserving SRs.
4. Day 4. Dilute cultures 1:1,000 and incubate at 30 °C overnight.
5. Day 5. Dilute cultures 1:1,000 and incubate at 30 °C overnight.
6. Day 6. Dilute cultures 1:1,000 and incubate at 30 °C overnight. In addition, dilution plate cultures at final dilutions of 10^{-7} and 10^{-8} on TSA and incubate overnight at 37 °C. *See* **Note 5** for counter-selection plasmids.
7. Day 7. Repeat Day 6. In addition, replica-patch colonies from Day 6 onto TSA and TSA with antibiotic. Incubate plates overnight at 37 °C.
8. Day 8. Check patches for antibiotic-sensitive colonies. These are double recombinants and should be screened by PCR to confirm mutation. If no or only several antibiotic-sensitive colonies or if PCR only indicates wild-type genotype, patch Day 7 dilution plating colonies (*see* **Note 6**).
9. Day 9. Only necessary if there are no antibiotic-sensitive colonies on Day 8 or if PCR only shows wild-type genotype. If needed, repeat Day8.
10. Following analysis of generated mutants, use other genetic tools outlined in **Note** 7 to confirm phenotypes observed with mutant.

4 Notes

1. Recombination occurs between homologous pieces of DNA and therefore the size of DNA fragment used will partially dictate the frequency of recombination. For efficient frequency, it is recommended to use 500–1,000 base pairs of flanking sequence for mutant making. To this end, for a point mutation, a 2-kb fragment centered on the mutation site is recommended. For a deletion plasmid, 500–1,000 bp upstream and downstream of the deletion site is suggested.

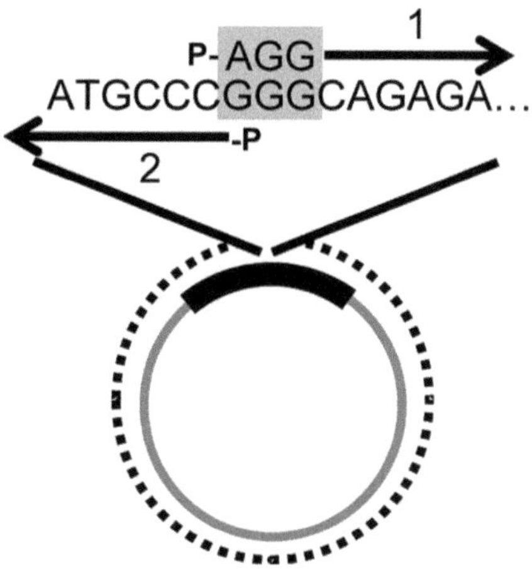

Fig. 2 Generation of point mutations. Two primers (1 and 2) are ordered in opposite orientations to PCR (*dotted line*) around the plasmid. Primer 1 contains the desired mutation GGG→AGG. Primers are purchased with 5′ phosphates so that following PCR, the product can be self-ligated. Following ligation, the sample is treated with DpnI which only cleaves methylated DNA to destroy the template. Note that mutation removes the native SmaI site (CCCGG) to facilitate screening

2. Generating mutations for an allelic exchange plasmid

   Generating point mutations. It is often advantageous to be able to produce small mutations such as the insertion of a stop codon to prematurely terminate a protein of interest or to make amino acid substitutions. The principle is the same whether the desire is to place the mutation on the chromosome or for use in *trans* complementation experiments using plasmids While several methods such as mutagenesis kits or PCR SOEing can be used, we prefer an alternate PCR-based procedure. For this, the target DNA is first placed into a small *E. coli* vector such as pCR-Blunt. If it is for chromosome integration, use PCR to amplify ~1 kb on either side of the desired mutation site. This 2-kb fragment is first moved into pCR-Blunt and verified by sequencing. Next, PCR primers are designed in which one of the primers contains the desired mutation close to the 5′ end of the oligonucleotide (Fig. 2). The primers are designed around the target site and face in opposite directions such that the PCR progresses away from the target site and around the plasmid. For example, as seen in Fig. 2, primer 1 start as 5′-AGGCAGAGA and contain the desired mutation (GGG to AGG) while primer 2 is complementary to the DNA flanking the mutation site and would start as 5′-GGGCAT (*see* **Note 8**). The mutation is therefore incorporated in the PCR product as the entire plasmid is amplified. Since this is a sizeable PCR (~5.5 kb) including both the 2-kb cloned fragment and 3.5-kb pCR-Blunt plasmid, it is

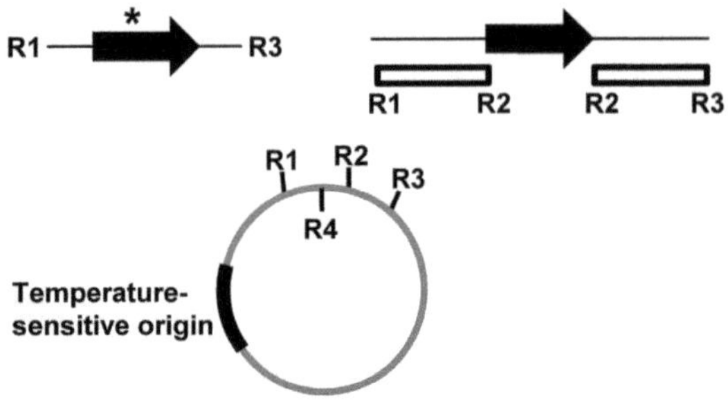

Fig. 3 Generation of allelic exchange plasmid. A point mutation (*left*) can be cloned into the temperature-sensitive plasmid using enzymes R1 and R3. For deletion plasmids, upstream and downstream flanking fragments are cloned into the plasmid using R1/R2 and R2/R3. Enzyme R4 can be used as a vector kill for either the point mutation or the upstream fragment

important to choose an appropriate polymerase. We prefer KOD polymerase because it has a low mutation frequency, high processivity, and a rapid elongation rate, requiring a PCR extension time as short as 45 s. Importantly, the PCR primers must be ordered with attached 5′ phosphorylation and is a standard modification available through most vendors. These phosphates are necessary because T4 DNA ligase requires a free 5′ phosphate group. Following confirmation of the PCR by agarose gel-electrophoresis, the PCR product is self-ligated using T4 DNA ligase to circularize the product into a plasmid. The tube now contains a mixture of the original template plasmid as well as the desired mutated version. The mix should be treated with the restriction endonuclease DpnI (3 h to overnight), which only recognizes methylated DNA, to remove the original template. This treated sample is then transformed into *E. coli* and resulting plasmids verified by sequencing. The mutation can then be placed in an appropriate vector for expression or allelic exchange (*see* Table 1).

Generating deletion mutants. Similar to generating point mutations, there are several techniques used to create deletion mutant plasmids. Many use PCR SOEing, but we will cover a simpler cloning technique. While this can be used to perform any size deletion, presented here is an example of a single gene complete deletion. In this procedure, fragments upstream and downstream of the target gene are cloned into the temperature-sensitive allelic exchange plasmid (Fig. 3). In the example, PCR is used to amplify 1,000 bp of upstream flanking with restriction sites for enzymes R1 and R2 and a downstream fragment with sites for R2 and R3. For a complete deletion, R2 in each fragment should be flanked by that start and stop codons of the gene, e.g., <u>ATG</u>GAATTC<u>TAA</u>, leaving the ATG and TAA in the same translation reading frame.

The upstream and downstream fragments can be cloned sequentially into the plasmid. For an alternative combined ligation step *see* **Note 9**. If a marked mutation is desired, an antibiotic resistance gene can be cloned into the deletion using R2.

Vector kill. Remnant original template or self-ligated, partially digested, parent plasmid can make it through any cloning procedure. This often appears as the original vector, i.e., no insert, identified during the screening process. To increase the odds of finding a successful clone with insert, a vector kill can be used. A vector kill is a restriction digest using a site that was removed during the cloning procedure. This usually can be accomplished in two ways. First, using an enzyme site removed from the plasmid. For example, cloning between EcoRI and NheI in a multiple cloning site with sequential EcoRI–SalI–NheI sites removes the SalI site. Therefore, any original vector can be removed by digesting the ligation reaction with SalI. Second, is using compatible ends to destroy a recognition site. For example, the overhangs following digestion with AvrII, NheI, XbaI, or SpeI can be ligated to each other. Following ligation, none of the original cut sites are retained. So cloning a fragment digested with EcoRI and AvrII into the example above would destroy the NheI site. In this case, one could vector kill with either SalI or NheI. For a good list of compatible blunt and cohesive termini, see the New England Biolabs catalog. Vector killing for several hours or overnight by either of these methods will increase the proportion of the clones containing insert.

3. For an unknown reason, *S. epidermidis* does not produce the large and small colonies during growth at 44 °C. Therefore, colonies that grow at this temperature need to be checked for plasmid integration. This can be accomplished by two means. First, PCR can be performed using a primer outside the cloned upstream and downstream regions. These primers are paired with an internal primer for the plasmid. A PCR for both upstream and downstream integrations should be checked since theoretically each should occur 50 % of the time. Secondly, integration can be confirmed by southern blot analysis using plasmid sequence as a probe.
4. It is a good idea to make −80 °C stocks of the SRs in case no mutants are found by the end of the procedure. In this case, a single colony is used to inoculate TSB with antibiotic and incubated at 44 °C until turbid. Doing so prevents the need of redoing the high-temperature streak plating a second time.
5. Some plasmids such as pKOR1 and pJB38 allow for the use of counter-selection. They encode an anhydrotetracycline-inducible promoter that makes an antisense *secY* transcript which inhibits cell growth. If these plasmids are used, plate overnight cultures starting on Day 6 at 10^{-7} final dilution on

TSA supplemented with 100 ng/ml anhydrotetracycline. The following day, patch the large colonies. While this system helps, it is not perfect and there will still be some large colonies with the plasmid.

6. Identifying double recombinants requires two events to occur, recombination of the plasmid out of the chromosome and loss of the plasmid (source of the antibiotic resistance). Since both of these events are theoretically random events, they can happen at any time during the nonselective 30 °C growth. If this happens early during the sequential subculturing, DRs will be seen earlier in the patch screening. Due to this, it is not uncommon to find zero DRs during the first day of patching, requiring a second or third day of dilution plating, i.e., plating on Day 7 and Day 8. In addition, while statistically there should be a result of DRs being 50 % wild type and 50 % mutant, there are times where an SR will only give one type of DR (wild type or mutant only). This is when it is convenient to have made stocks of multiple SRs (*see* **Note 4**). Finally, if the recombination and plasmid loss occurs early enough in the subculturing steps, that DR could overpopulate the culture and yield an unusually high number of DRs on the first day of patching. These are certainly all clones of each other and therefore only several DRs should be screened and the culture no longer dilution plated if they are all of the wild-type genotype.

7. Other genetic tools for characterizing mutants.

   Complementation. Following the generation of mutants, genetic complementation is necessary to ensure that observed phenotypes are due to the generated mutation and not the result of secondary-site mutations. There are three types of genetic complementation to consider. First, is *trans* complementation on a plasmid. This is the most convenient and common form of complementation and involves expression of the wild-type allele from a plasmid. There are a multitude of plasmids available for these studies, a few of which are suggested in Table 1. Important considerations when choosing a plasmid include: antibiotic selection marker, copy number, plasmid-incompatibility, and plasmid stability without selection. Secondly, *trans* complementation can be performed by inserting the wild-type allele at an alternate location on the chromosome. The benefit to this method is that complementation is in single-copy and does not require antibiotic selection. However, it does require a secondary site mutation be generated, i.e., the location of the *trans* complementation insertion. This can be performed simply using allelic exchange. Importantly, either of these methods may not complement perfectly and may under- or over-complement the phenotype. Finally, there are times where *trans* complementation does not work or cannot be performed. In those cases, the wild-type allele should be

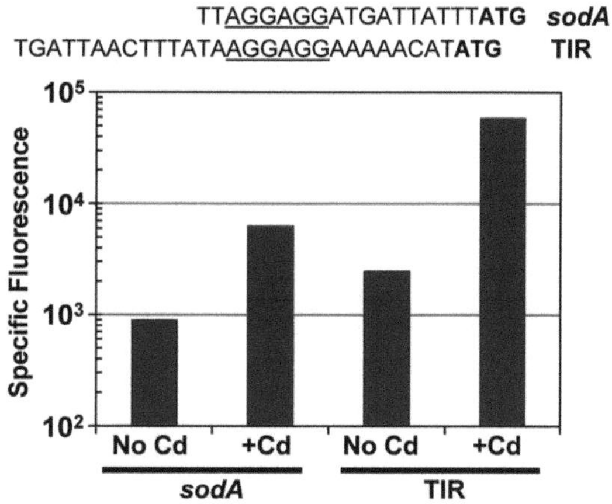

Fig. 4 Enhanced expression by TIR compared to the *sodA* RBS. Fluorescence of GFP under control of P_{cad} in pCN51 with or without 5 μM cadmium induction. Note that there is increased fluorescence from TIR despite having the same Shine–Dalgarno sequence AGGAGG

restored back onto the chromosome where it originally existed. This is sometimes called reverse complementation.

There are several choices that need to be made when choosing an expression system. Obviously, plasmids have to be chosen with appropriate antibiotic resistance markers. A second consideration is plasmid replication origin (oriV) which determines the copy number, stability (maintenance without selection), and compatibility with native or other transformed plasmids. The next decision is that of how the gene of interest is expressed. While it is often best to use the native promoter and ribosome binding site (RBS), at times a controlled or constitutive expression may be desired. There are several inducible systems available such as P_{cad} in pCN51 [1] and $P_{xyl/tetO}$ [2] which respond to cadmium (5–10 μM) and anhydrotetracycline (100–250 ng/ml), respectively, while the promoter for *sarA* is a good candidate for constitutive expression [3]. To increase expression for proteins due to a poor RBS an alternate RBS can be used. For optimal expression, we prefer the TIR (*T*ranslation *I*nitiation *R*egion) which contains a consensus Shine–Dalgarno sequence with optimal spacing to the start codon and the Phage T7 enhancer region to minimize secondary structure [4, 5]. As shown in Fig. 4, the TIR leads to a greater induction than the *sodA* RBS (used for enhanced expression) despite having the same Shine–Dalgarno site (AGGAGG).

Gene expression using reporter proteins. In addition to characterizing a mutant, it is often necessary to understand the expression of a gene of interest. One way of doing so is through the use of transcriptional and translation reporters. There are a myriad of reporter genes available and one should be picked for the desired application. The *lacZ*-encoded enzyme β-galactosidase has long been used as a reporter because it is easily detected with colorimetric substrates such as ortho-Nitrophenyl-β-galactoside (ONPG) and 5-bromo-4-chloro-3-indolyl-β-D-galactopyranoside (X-gal) as well as the fluorescent substrates methylumbelliferone β-D-galactopyranoside (MUG) and fluorescein di-β-D-galactopyranoside (FDG). In addition, the units of the β-galactosidase assay are well described [6]. However, β-galactosidase is most often used on supernatants and/or cell extracts and can be difficult to clone due to its ~3 kb size. For live cells, most people choose to express a fluorescent proteins such as GFP and its derivatives. There are a variety of colors available with excitation and emission wavelengths suiting most flow cytometry facilities. In addition, they can be easily monitored using epifluorescent and confocal microscopy as well as by fluorescent readers. Typically these proteins are extremely stable and therefore give a representation of accumulated expression and are only detected upon reaching a threshold concentration. However, there are short half-life versions available. We have recently generated a selection of fluorescent proteins codon-optimized specifically for staphylococci [7]. Bioluminescence via the *lux* genes has also been developed [1, 8]; however, staphylococci do not produce significant levels of the Lux aldehyde substrate and therefore require the addition of aldehyde-synthesis proteins [8]. Finally, there are other less used reporters such as *xylE* (catechol 2,3-dioxygenase) [9], *blaZ* (β-lactamase) [10], and *cat* (chloramphenicol acetyltransferase) [11]. All of these genes can be expressed from a plasmid or integrated into the chromosome using allelic exchange.

Transcriptional reporters are used to monitor the activity of promoters. With this is mind, often nonnative RBSs can used to produce a better signal for detection and monitoring purposes. For example, to examine the expression from the promoter controlling the *xyz* operon (P_{XYZ}), the promoter region including all regulatory elements and the transcription initiation site should be cloned into a plasmid. The TIR attached the *gfp* which encodes GFP is then placed downstream of P_{XYZ}. Thus, fluorescence is the output that indicates when and to what levels P_{XYZ} is active. This activity is then measured quantitatively with fluorescent reader or qualitatively by microscopy.

Much like transcriptional reporters, translation reporters use the promoter of the target gene but include the native RBS and 5′ untranslated region (UTR). In this way, the expression of the reporter best mimics the expression of the protein of interest. However, due to the nature of these reporters it is not possible to amplify the signal using optimal promoter and RBS elements. In this case, the promoter (including all regulatory elements) and the 5′ UTR which includes the RBS and all sequences including the protein start codon are cloned into a plasmid. If desired, more than the start codon can be included. It is important to include a restriction cut site after the start codon. A reporter gene is then cloned in-frame with the start codon of the target protein, i.e., ensure that the reading frame of the reporter gene aligns with the target protein's start codon. This allows expression from the promoter to read through the native RBS and start codon and into the gene encoding the reporter protein.

8. If possible, the point mutation should either remove or generate a restriction endonuclease recognition site. In this case, the example removes a SmaI site (CCCGGG). This makes screening for the desired mutation on the chromosome a lot easier since one can simple digest a PCR product to confirm that a double recombinant has the wild-type or mutant allele. This is often not a possibility and therefore sequencing of a PCR product may be the only screen available.
9. We have been having success with a single-step ligation procedure with all three DNA fragments. The plasmid is digested with R1 and R3, the upstream PCR product is cut with R1 and R2, and the downstream fragment is digested with R2 and R3. All three fragments are mixed in a single ligation step with equal amounts of upstream and downstream fragments as visualized on a gel.

References

1. Charpentier E, Anton AI, Barry P et al (2004) Novel cassette-based shuttle vector system for gram-positive bacteria. Appl Environ Microbiol 70:6076–6085
2. Bateman BT, Donegan NP, Jarry TM et al (2001) Evaluation of a tetracycline-inducible promoter in *Staphylococcus aureus* in vitro and in vivo and its application in demonstrating the role of sigB in microcolony formation. Infect Immun 69:7851–7857
3. Malone CL, Boles BR, Lauderdale KJ et al (2009) Fluorescent reporters for *Staphylococcus aureus*. J Microbiol Methods 77:251–260
4. Cheng X, Patterson TA (1992) Construction and use of lambda PL promoter vectors for direct cloning and high level expression of PCR amplified DNA coding sequences. Nucleic Acids Res 20:4591–4598
5. Miller WG, Lindow SE (1997) An improved GFP cloning cassette designed for prokaryotic transcriptional fusions. Gene 191:149–153
6. Miller JH (1992) A short course in bacterial genetics. Cold Spring Harbor Laboratory Press, Cold Spring Harbor, NY
7. Bose JL, Fey PD, Bayles KW (2013) Genetic tools to enhance the study of gene function and regulation in *Staphylococcus aureus*. Appl Environ Microbiol 79:2218–2224
8. Mesak LR, Yim G, Davies J (2009) Improved *lux* reporters for use in *Staphylococcus aureus*. Plasmid 61:182–187
9. Sheehan BJ, Foster TJ, Dorman CJ et al (1992) Osmotic and growth-phase dependent regulation of the eta gene of *Staphylococcus aureus*: a role for DNA supercoiling. Mol Gen Genet 232:49–57
10. Grkovic S, Brown MH, Hardie KM et al (2003) Stable low-copy-number *Staphylococcus aureus* shuttle vectors. Microbiology 149: 785–794
11. Kwong SM, Skurray RA, Firth N (2004) *Staphylococcus aureus* multiresistance plasmid pSK41: analysis of the replication region, initiator protein binding and antisense RNA regulation. Mol Microbiol 51:497–509
12. Sau S, Sun J, Lee CY (1997) Molecular characterization and transcriptional analysis of type 8 capsule genes in *Staphylococcus aureus*. J Bacteriol 179:1614–1621
13. Bae T, Schneewind O (2006) Allelic replacement in *Staphylococcus aureus* with inducible counter-selection. Plasmid 55:58–63
14. Lee CY, Buranen SL, Ye ZH (1991) Construction of single-copy integration vectors for *Staphylococcus aureus*. Gene 103: 101–105

Chapter 9

Isolation of Chromosomal and Plasmid DNA from *Staphylococcus epidermidis*

Jill K. Lindgren

Abstract

The following describes noncommercial methods for the purification of genomic and plasmid DNA from *S. epidermidis*. These include both large-scale, high molecular weight and quick, small-scale chromosomal DNA extractions, and also a standard alkaline lysis method of plasmid preparation.

Key words *Staphylococcus epidermidis*, Chromosomal DNA, Plasmid DNA, Lysostaphin

1 Introduction

There are several noncommercial protocols available for the isolation of DNA from *S. epidermidis* (*see* **Note 1**). In most cases, standard laboratory methods for nucleic acid isolation are appropriate for use with *S. epidermidis* with slight modification, as the staphylococcal cell wall is resistant to lysis by standard lysozyme and/or detergent methods. Efficient cell lysis can be achieved using either mechanical disruption of the cell wall via bead beating, or enzymatic digestion of the cell wall via lysostaphin. Lysostaphin is a commercially available endopeptidase which specifically cleaves between glycine residues in the pentaglycine cross-bridge of *Staphylococcus aureus* peptidoglycan [1]. Due to the presence of serine residues in the pentapeptide cross-bridges of *S. epidermidis* peptidoglycan, a higher concentration of lysostaphin and longer incubation time are required for complete cell lysis [2, 3]. Lysostaphin is the preferred method of cell lysis for plasmid or high molecular weight, genomic DNA, as mechanical lysis may cause shearing of chromosomal DNA.

Use of the following methods should generate chromosomal and plasmid DNA with sufficient quality and quantity for most laboratory applications. Subheading 3.1 describes a large-scale chromosomal DNA extraction. This protocol is based on the

Paul D. Fey (ed.), *Staphylococcus Epidermidis: Methods and Protocols*, Methods in Molecular Biology, vol. 1106, DOI 10.1007/978-1-62703-736-5_9, © Springer Science+Business Media, LLC 2014

method reported by Marmur in 1961 and generates a large yield of high molecular weight DNA [4]. This method minimizes DNA shearing and is therefore useful when high-quality DNA is necessary, such as the generation of a DNA library for whole genome sequencing. Subheading 3.2 describes a small-scale preparation of chromosomal DNA. This relatively quick protocol will generate about 2–5 μg of genomic DNA and can be used for PCR template, sequence typing, Southern blotting, etc. Subheading 3.3 describes a traditional alkaline lysis preparation of plasmid DNA from *S. epidermidis* [5]. This method typically yields approximately 500 ng/μl of plasmid DNA suitable for PCR, cloning, transformation, etc. These methods provide reliable, lower cost alternatives to commercial DNA isolation kits.

2 Materials

2.1 Reagents for Large-Scale Chromosomal DNA Extraction

1. TE buffer: 10 mM Tris–HCl, 1 mM EDTA, pH 7.6.
2. TES buffer: 50 mM Tris–HCl, 5 mM EDTA, 50 mM sodium chloride.
3. Lysostaphin (5 mg/ml).
4. RNase A (10 mg/ml).
5. Proteinase K (10 mg/ml).
6. 20 % SDS.
7. 5 M Sodium perchlorate.
8. Chloroform solution: chloroform:isoamyl alcohol 24:1.
9. 95 % Ethanol (4 °C).
10. 1× SSC buffer: 15 mM sodium citrate, 150 mM sodium chloride. To prepare a stock solution of 10× SSC, dissolve 4.41 g of sodium citrate and 8.77 g of sodium chloride in 80 ml of deionized water. Titrate pH to 7.2 with HCl and adjust final volume to 100 ml with water. Dilute tenfold prior to use.

2.2 Reagents for Minipreparation of Chromosomal DNA

1. CTAB–NaCl: 10 % hexadecyltrimethylammonium bromide (CTAB) in 0.7 M NaCl (*see* **Note 10**).
2. TE buffer: 10 mM Tris–HCl, 1 mM EDTA.
3. Lysostaphin (1 mg/ml).
4. 10 % SDS.
5. Proteinase K (20 mg/ml).
6. 5 M NaCl.
7. Phenol:chloroform solution: saturated phenol:chloroform: isoamyl alcohol 25:24:1.

8. Chloroform solution: chloroform:isoamyl alcohol 24:1.
9. Isopropanol (4 °C).
10. 70 % Ethanol (4 °C).

2.3 Reagents for Alkaline Lysis Preparation of Plasmid DNA

1. 3 % Sodium chloride solution (w/v).
2. Solution I (TES): 20 mM Tris–HCl, pH 7.0, 50 mM EDTA, pH 8.0, 0.58 M sucrose.
3. Lysostaphin (1 mg/ml).
4. Solution II: 0.1 M NaOH, 1 % SDS (*see* **Note 11**).
5. Solution III: 1.5 M potassium acetate, pH 4.8.
6. Phenol:chloroform:isoamyl alcohol 25:24:1.
7. Ethanol: 100 and 70 %.
8. TE buffer: 10 mM Tris–HCl, pH 7.5, 1 mM EDTA, pH 8.0.

3 Methods

3.1 Large-Scale Chromosomal DNA Extraction [4]

1. Harvest *S. epidermidis* cells from a 150 ml overnight culture grown in rich media (*see* **Note 2**) by centrifugation for 10 min at 4,000 × *g*; 4 °C. Discard the supernatant.
2. Wash the cell pellet by resuspension in 20 ml of TE buffer.
3. Centrifuge cells as in **step 1** and discard the supernatant.
4. Resuspend the cell pellet in 5 ml of TES buffer and transfer to a 125 ml flask.
5. Add 250 μl of lysostaphin and 50 μl of RNase A. Incubate at 37 °C with agitation at a low RPM for 30 min or until lysis appears complete (*see* **Note 3**).
6. Add 50 μl of proteinase K and 125 μl of SDS. Incubate at 37 °C with agitation at a low RPM for 1 h.
7. Add 7 ml of 5 M sodium perchlorate and mix by gentle swirling until the cellular debris is cleared (*see* **Note 4**).
8. Add 30 ml of chloroform:isoamyl alcohol and mix by gentle swirling. Place on an ice bath agitator at low RPM for 15 min (*see* **Note 5**).
9. Divide the mixture into two 30 ml chloroform-resistant centrifuge tubes and centrifuge for 45 min at 7,600 × *g* at 4 °C.
10. Carefully transfer the aqueous phase (top layer) to a sterile 125 ml flask using a wide mouth pipette to prevent shearing of DNA. Discard the organic phase (*see* **Note 6**). If cellular debris remains in the aqueous phase, repeat **steps 8–10**. If the aqueous phase is clear proceed to **step 11**.
11. Slowly (*see* **Note 7**) overlay 20 ml of cold 95 % ethanol onto the aqueous phase. Chromosomal DNA will precipitate at the interphase.

12. The resulting DNA can be spooled on a sterilized, glass Pasteur pipette and transferred to a tube containing 2 ml of TE buffer (*see* **Note 8**).
13. Incubate overnight at 4 °C to allow the DNA to fully dissolve (*see* **Note 9**).
14. If necessary, the DNA can be dialyzed against 2 l of TE or desired buffer overnight at 4 °C with one exchange.
15. DNA can be stored at 4 °C short-term, or −20 °C for longer storage.

3.2 Minipreparation of Chromosomal DNA

1. Centrifuge 3 ml of overnight culture of *S. epidermidis* at 6,000 × *g* for 10 min (*see* **Note 2**).
2. Discard the supernatant and resuspend the cell pellet in 500 μl of TE buffer.
3. Add 25 μl of lysostaphin (1 mg/ml). Mix gently by inversion and incubate at 37 °C for 60 min or until the cells are well lysed (*see* **Notes 3** and **4**).
4. Add 30 μl of 10 % SDS.
5. Add 3 μl of proteinase K (20 mg/ml).
6. Mix gently by inversion and incubate at 37°°C for 15–30 min.
7. Add 100 μl of 5 M NaCl and mix gently by inversion.
8. Add 80 μl of CTAB–NaCl solution and mix gently by inversion (*see* **Note 10**).
9. Incubate at 65 °C for 10 min.
10. Add an equal volume of phenol:chloroform solution and mix by inversion (*see* **Note 5**).
11. Centrifuge 5 min at maximum speed (>10,000 × *g*) to separate phases.
12. Transfer the aqueous phase (top layer) to a clean tube and discard the organic phase (*see* **Note 6**).
13. Add an equal volume of chloroform solution and mix by inversion.
14. Centrifuge 1 min at maximum speed.
15. Transfer the aqueous phase to a clean tube and discard the organic phase.
16. Add 0.6 volumes of cold isopropanol and mix by inversion.
17. Centrifuge for 15 min at maximum speed and discard the supernatant.
18. Wash the DNA pellet with cold 70 % ethanol.
19. Centrifuge 15 min at maximum speed and discard ethanol.
20. Dry DNA pellet slightly and resuspend in 50 μl of desired elution buffer (TE, water, etc.).

3.3 Alkaline Lysis Preparation of Plasmid DNA [5]

1. Pellet cells from 2–10 ml of overnight culture by centrifugation for 5–10 min at 6,000 × *g* at room temperature (*see* **Note 2**). Discard the supernatant.
2. Wash the cells by resuspension in 1 ml of 3 % sodium chloride solution.
3. Pellet the cells by centrifugation at 6,000 × *g* at room temperature for 5–10 min and discard the supernatant.
4. Resuspend the cell pellet in 150 μl of Solution I.
5. Add 10 μl of lysostaphin and incubate at 37 °C for 30 min or until lysis is complete (*see* **Notes 3** and **4**).
6. Add 300 μl of Solution II (*see* **Note 11**) and mix by gentle inversion. Incubate on ice for 5 min.
7. Add 225 μl of Solution III and mix by gentle inversion. Incubate on ice for 5 min.
8. Centrifuge the lysate at maximum speed (>10,000 × *g*) at room temperature for 10 min and transfer the supernatant to a new tube (*see* **Note 12**).
9. Add 500 μl of phenol:chloroform solution and vortex to mix well (*see* **Notes 5** and **13**).
10. Centrifuge at maximum speed for 5 min to separate phases.
11. Transfer the aqueous phase (top layer) to a clean centrifuge tube. Discard the organic phase (*see* **Note 6**).
12. Add two volumes of cold 100 % ethanol and freeze at −80 °C for 20 min (or −20 °C overnight) to precipitate the DNA.
13. Centrifuge at maximum speed for 15 min at room temperature and discard supernatant.
14. Wash the DNA pellet by addition of 1 ml of 70 % ethanol.
15. Centrifuge at maximum speed for 15 min at room temperature and discard supernatant.
16. Dissolve DNA pellet in 40 μl of desired diluent; i.e., TE buffer or water.

4 Notes

1. Commercially available kits can be used successfully for nucleic acid preparation from *S. epidermidis*. Either mechanical cell disruption or incubation with lysostaphin (100 μg/ml) at 37 °C for 30–60 min prior to addition of the commercial lysis buffer will ensure complete cell lysis.
2. Overnight cultures should be inoculated from a single colony from a freshly streaked plate and grown in rich media (tryptic soy broth, brain heart infusion broth, etc.) at 37 °C with shaking at 250 rpm.

3. As cells are lysed, the mixture will become cloudy and take on a more viscous, "snotty" appearance. Complete lysis of the cell pellet is necessary for obtaining a high yield of DNA.
4. Handle the sample gently from this point to prevent shearing of chromosomal DNA.
5. Steps involving phenol and/or chloroform should be performed in a fume hood and caution should be taken.
6. Organic waste must be disposed of according to federal and institutional regulations.
7. It is important that the ethanol is poured slowly down the side of the flask so that it overlays the aqueous phase rather than mixing with it. As you swirl the DNA on the glass pipette, you will see the DNA spooling onto the rod.
8. If TE is incompatible with downstream applications, an alternate elution buffer, such as water, can be used.
9. If the DNA remains bound to the glass pipette, incubate the pipette in the buffer overnight and allow the DNA to dissolve into solution.
10. Dissolve the sodium chloride in water and slowly add the CTAB while stirring and heating. Heat the solution to 65 °C if necessary and adjust to final volume. Before use, CTAB–NaCl should be warmed to 55 °C to aid in pipetting.
11. Prepare fresh solution II for each use. It is convenient to prepare stocks of 1 M NaOH and 10 % SDS, which can be diluted tenfold immediately before use.
12. If any cellular debris remains, repeat centrifugation to clarify.
13. Phenol:chloroform extraction of plasmid DNA is an optional step, but may be necessary for downstream applications requiring higher purity DNA.

References

1. Browder HP, Zygmunt WA, Young JR et al (1965) Lysostaphin: enzymatic mode of action. Biochem Biophys Res Commun 19: 383–389
2. Wu JA, Kusuma C, Mond JJ et al (2003) Lysostaphin disrupts *Staphylococcus aureus* and *Staphylococcus epidermidis* biofilms on artificial surfaces. Antimicrob Agents Chemother 47: 3407–3414
3. Tipper DJ, Berman MF (1969) Structures of the cell wall peptidoglycans of *Staphylococcus epidermidis* Texas 26 and *Staphylococcus aureus* Copenhagen.I. Chain length and average sequence of cross-bridge peptides. Biochemistry 8:2183–2192
4. Marmur J (1961) A procedure for the isolation of deoxyribonucleic acid from micro-organisms. J Mol Biol 3:208–218
5. MacNamara PJ (2008) Genetic manipulation in *Staphylococcus aureus*. In: Lindsay J (ed) *Staphylococcus*: molecular genetics. Caister Academic Press, Norfolk, UK, pp 89–129

Chapter 10

Isolation of *Staphylococcus* sp. RNA

Tess Eidem

Abstract

Isolation of RNA (ribonucleic acid) is a valuable technique to study gene regulation and functional RNAs. It is important to obtain pure samples of RNA for downstream applications, while avoiding the negative effects of ribonucleases (RNases). Here we describe several methods of extracting RNA, with an emphasis on *Staphylococcus* RNA isolation, and measures that must be taken to ensure high-quality, high-yield RNA.

Key words RNA purification of Gram-positive bacteria, *Staphylococcus* gene expression, RNase prevention, Phenol–chloroform free, Silica column, RNA quality control

1 Introduction

RNA (ribonucleic acid) is a macromolecule found in all three kingdoms of life. RNA polymers are made of ribonucleotides connected by a phosphodiester bond, and each ribonucleotide contains a pentose sugar, a phosphate group, and a nitrogenous base. While the structurally similar DNA encodes an organism's genes, RNA plays a large role in determining how those genes are expressed. In addition to being a "messenger" for protein expression, RNA can also be functional as tRNAs, rRNAs, small regulatory RNAs, and catalytic ribozymes. All cells ultimately function as a product of their gene expression profile, therefore analysis of RNA modulation allows a researcher to gain valuable insight on many cellular processes [1].

Ribonucleases (RNases) are enzymes that degrade RNA. Due to RNase abundance in dust and on the human skin, combined with their extremely stable nature, gloves must always be worn when handling samples, all pipets and equipment should be cleaned with RNaseZap (Ambion AM9780) or equivalent, pipet tips and other plasticware or glassware must be autoclaved, and buffers and water should be designated as nuclease-free or treated with DEPC (diethyl pyrocarbonate) and autoclaved before use. The greatest barrier to RNA isolation in Gram-positive bacteria is

Paul D. Fey (ed.), *Staphylococcus Epidermidis: Methods and Protocols*, Methods in Molecular Biology, vol. 1106, DOI 10.1007/978-1-62703-736-5_10, © Springer Science+Business Media, LLC 2014

the thick peptidoglycan cell wall. Breaching the cell wall involves mechanical disruption, such as homogenization using a tissue homogenizer as discussed later, or breaking apart the cell wall by enzymatic digestion, as with lysostaphin. QIAGEN recommends extracting RNA from exponentially growing bacteria, as stationary-phase cell walls are more difficult to disrupt for RNA isolation [2]. Although not discussed in full detail here, there are recent improved methods for isolating RNA from *Staphylococcus* biofilms [3, 4].

As RNA turnover occurs rapidly in bacteria (average mRNA half-life ≤2.0 min), it is essential that cell lysis and stabilization of RNA occur rapidly [5]. To ensure RNA integrity, endogenous RNases within the sample must be inactivated. This can be performed through the addition of denaturing salts, organic solutions, detergents, reducing agents, or a variety of enzymes that eliminate or block RNase activity [1]. After cell wall disruption, there are several methods by which RNA can be isolated, commonly by acid–phenol–chloroform extraction [6]. This approach allows for simultaneous inactivation of RNases, as well as separation of RNA from other macromolecules. Resulting cell lysates are resuspended in an acidic (pH ~4.8) phenol mixture (commonly containing guanidine thiocyanate, a chaotropic salt) followed by addition of chloroform, causing an organic phase, interphase, and aqueous phase to form. The organic phase and the interphase contain lipids, proteins, and DNA, while RNA partitions into the aqueous phase. The aqueous phase can then be removed, the RNA precipitated with various combinations of salt and alcohol, followed by several washes to remove contaminating salts, as well as residual phenol and chloroform.

Although this extraction method is widely used, it has several shortcomings. Phenol is a hazardous chemical posing potential health risks, and its carryover from the RNA extraction process can skew the A_{260}/A_{280} ratio (*see* **Note 16**) and interfere with downstream enzymatic applications, such as reverse transcriptase [7]. As such, our laboratory utilizes the phenol-free QIAGEN RNeasy Kit, which extracts RNA using guanidine thiocyanate, binds and washes the RNA on manufactured silica columns, and results in high-quality, high-yield RNA [8, 9]. There are kit-free methods to isolate RNA from *Staphylococcus* species; however, our laboratory favors the use of RNA isolation kits due to their reproducibility, excellent RNA quality, and high RNA yield.

2 Materials

2.1 RNA Extraction Using the QIAGEN RNeasy Kit

1. Acetone:ethanol (1:1 vol/vol; *see* **Note 1**).
2. β-Mercaptoethanol.
3. TE Buffer (10 mM Tris–HCl, 1 mM EDTA, pH 8).

4. Ready-to-use 2-ml tubes containing 0.1 mm silica spheres, Lysing Matrix B (MP Biomedicals 116911050).
5. FastPrep-24 tissue homogenizer (MP Biomedicals 116004500).
6. Qiagen RNeasy Kit (QIAGEN 74104; *see* **Notes 2–6**).
7. Microcentrifuge.
8. Benchtop centrifuge.
9. Sterile, nuclease-free microcentrifuge tubes.
10. Sterile, nuclease-free pipet tips.
11. Nuclease-free water.
12. Disposable gloves (*see* **Note 7**).
13. −80 °C freezer.

3 Methods

3.1 RNA Extraction Using the QIAGEN RNeasy Kit

1. Harvest bacterial cells by removing 4–20 ml of exponential-phase cultures and immediately adding an equal volume of ice-cold acetone:ethanol (1:1 vol/vol). Mix thoroughly by vortexing (*see* **Note 8**).
2. Centrifuge at 1,510 × *g* for 10 min at 4 °C to pellet the cells.
3. Pour off supernatant and resuspend pellet in 500 μl ice-cold TE buffer (*see* **Note 9**).
4. Place sample into 2-ml Lysis Matrix B tube containing 0.1 mm silica spheres (MP Biomedicals).
5. Homogenize sample in the MP Biomedicals tissue homogenizer at 5 m/s for 20 s.
6. Place samples on ice for 5 min.
7. Homogenize sample at 4.5 m/s for 20 s.
8. Centrifuge tubes for 15 min at 17,000 × *g* at 4 °C to remove cell debris.
9. Remove 200 μl of the cell lysate supernatant, avoiding the lysing matrix, and place into a new tube.
10. Immediately add 700 μl of buffer RLT containing β-mercaptoethanol (β-ME) to the lysate and mix well by pipetting (*see* **Note 10**). All subsequent steps can be performed at room temperature.
11. Add 500 μl 100 % ethanol to the sample and vortex.
12. Transfer 700 μl of the sample, including any precipitate that forms, to the RNeasy Mini spin column placed in a 2-ml collection tube (*see* **Note 11**).
13. Centrifuge sample in a microcentrifuge for 15 s at 17,000 × *g* and discard flow-through.

14. Place the remainder of the sample through the same spin column, centrifuge for 15 s at 17,000 × *g* and discard flow-through.
15. Add 700 μl Buffer RW1 to the column and centrifuge for 15 s at 17,000 × *g*. Discard the flow-through (*see* **Note 12**).
16. Add 500 μl Buffer RPE to the column and centrifuge for 15 s at 17,000 × *g*. Discard the flow-through (*see* **Note 13**).
17. Add an additional 500 μl Buffer RPE to the column and centrifuge for 15 s at 17,000 × *g*, discarding the flow-through.
18. Centrifuge the RNeasy spin column at 17,000 × *g* for 1 min to completely dry the membrane.
19. Place the RNeasy spin column into a new, appropriately labeled microcentrifuge tube.
20. Add 40 μl of nuclease-free water directly to the membrane and allow it to sit for 1 min, then centrifuge at 17,000 × *g* for 1 min to elute the RNA.
21. Remove the RNeasy column and discard.
22. Place the tube containing eluted RNA on ice or store at −80 °C (*see* **Note 14**).
23. Check RNA for concentration and quality (*see* **Notes 15–17**).
24. For downstream applications in which RNA samples must be free of DNA, such as qRT-PCR, we recommend an off-column DNase treatment (*see* **Note 18**).

4 Notes

1. Mix 100 % ethanol and 100 % acetone at a 1:1 vol/vol ratio. Store at −20 °C.
2. The RNeasy Mini spin column RNA binding capacity is 100 μg. The manufacturer recommends using no more than 1×10^9 bacteria per RNeasy Mini column. Overloading the column will significantly reduce the RNA yield and purity.
3. For assays requiring a larger number of cells, use the RNeasy Midi or RNeasy Maxi columns.
4. If interested in small RNA molecules (<200 nucleotides), we recommend using the miRNeasy Kit [10], as the RNeasy Kit does not yield high concentrations of small RNAs.
5. Add 10 μl β-mercaptoethanol (β-ME) per 1 ml of Buffer RLT and mix.
6. Buffer RPE is supplied as a concentrate. To obtain a working solution, add four volumes of 100 % ethanol, as indicated on the bottle.

7. Always wear gloves when handling RNA samples and during RNA isolation.
8. This step dehydrates the cells, halts RNA synthesis, and inactivates endogenous RNases. For stationary-phase cells, use no more than 4 ml of bacterial culture.
9. Stationary-phase cells require washing and more stringent lysing conditions. Thus, resuspend stationary-phase cell pellet in 1.5 ml ice-cold TE buffer, then centrifuge at 1,510 × *g* at 4 °C for 10 min and discard supernatant. Resuspend washed pellet in 500 μl of buffer RLT containing β-ME.
10. Buffer RLT contains a high concentration of guanidine isothiocyanate, a strong ionic protein denaturant, which encourages downstream RNA binding to the silica membrane. β-ME is a reducing agent that breaks disulfide bonds commonly found in RNases.
11. The RNeasy Mini spin column contains a silica matrix that binds the negatively charged RNA molecules.
12. Buffer RW1 is a stringent wash buffer that contains a guanidine salt and ethanol. This step removes small RNA molecules (<200 nucleotides) as well as carbohydrates, proteins, and fatty acids that are nonspecifically bound to the silica membrane.
13. Buffer RPE is a mild washing buffer used to remove the salts contained in the previous lysis and wash buffers. Ethanol is added to increase RNA binding to the silica membrane.
14. As RNA is naturally unstable, always keep purified RNA on ice for short-term use or store at −80 °C for up to a year.
15. Using the described RNA isolation protocol, RNA yields should routinely be 100–1,000 ng/μl.
16. The "260 to 280 ratio" (A_{260}/A_{280}) of a pure sample of RNA is 2.0 ± 0.15. Excess salt, DNA, and contaminating proteins can skew the absorbance of RNA, resulting in values outside of this range and most likely require additional extractions, washes, or cleanup steps [1].
17. RNA quality can also be assessed using gel electrophoresis using a denaturing formaldehyde agarose gel (1.2 % agarose, 0.66 M formaldehyde) followed by staining with ethidium bromide. Approximately 1 μg RNA is sufficient to visualize two clear bands representing the bacterial 16S and 23S rRNAs; smearing or absence of these bands indicates RNA degradation.
18. To remove any contaminating DNA, treat 1 μg RNA with DNase I (Ambion AM2222) according to the manufacturer's instructions, followed by an RNA cleanup using the RNeasy Mini Spin Columns.

References

1. Farrell REJ (2010) RNA methodologies: a laboratory guide for isolation and characterization. Academic, Burlington, MA
2. Qiagen (2005) RNAprotect® Bacteria Reagent Handbook. Qiagen, Valencia, CA
3. Atshan SS, Shamsudin MN, Lung LT et al (2012) Improved method for the isolation of RNA from bacteria refractory to disruption, including *S. aureus* producing biofilm. Gene 494:219–224
4. Franca A, Melo LD, Cerca N (2011) Comparison of RNA extraction methods from biofilm samples of *Staphylococcus epidermidis*. BMC Res Notes 4:572
5. Kushner SR (2004) mRNA decay in prokaryotes and eukaryotes: different approaches to a similar problem. IUBMB Life 56:585–594
6. Chomczynski P, Sacchi N (1987) Single-step method of RNA isolation by acid guanidinium thiocyanate-phenol-chloroform extraction. Anal Biochem 162:156–159
7. Gerstein AS (2001) Molecular biology problem solver: a laboratory guide. Wiley, New York
8. Qiagen (2010) RNeasy® Mini Handbook. Qiagen, Valencia, CA
9. Roberts C, Anderson KL, Murphy E et al (2006) Characterizing the effect of the *Staphylococcus aureus* virulence factor regulator, SarA, on log-phase mRNA half-lives. J Bacteriol 188:2593–2603
10. Qiagen (2012) miRNeasy Mini Handbook. Qiagen, Valencia, CA

Chapter 11

Use of Electroporation and Conjugative Mobilization for Genetic Manipulation of *Staphylococcus epidermidis*

Katherine L. Maliszewski and Austin S. Nuxoll

Abstract

To perform mechanistic studies on the biology of bacteria including metabolism, physiology, and pathogenesis, it is essential to possess the tools required for genetic manipulation. Introduction of plasmid DNA into *Staphylococcus epidermidis* for subsequent genetic manipulation, including allelic replacement and complementation experiments, is typically performed by either electroporation or conjugative mobilization. Herein, standard protocols and tips for the transfer of plasmid DNA to *S. epidermidis* by these two methods are provided.

Key words *Staphylococcus epidermidis*, *Staphylococcus*, Transformation, Competence, Electroporation, Conjugative mobilization, Conjugation

1 Introduction

Bacterial transformation is the process through which a recipient cell acquires DNA from its surrounding environment and integrates, through homologous recombination, novel genetic material into its chromosome. Along with phage transduction and conjugation, it is one of the major mechanisms of generating genetic diversity in bacterial populations. First described by Griffith in 1928 [1], DNA uptake occurs in a specialized physiological state termed competence. Since Griffith's experiments, induction of competence has been described in dozens of phylogenetically diverse species under widely varying conditions. Although the DNA binding and uptake apparatus (encoded by *com* genes) is highly conserved among competent species, the genes regulating the induction of a competent state are divergent.

The question of whether staphylococci are naturally competent has been ongoing for decades. With the increasing availability of sequenced *S. aureus* and *S. epidermidis* genomes, the presence of highly conserved *com* gene orthologs was noted in both species. However, despite the presence of orthologous *com* genes, *S. aureus*

Paul D. Fey (ed.), *Staphylococcus Epidermidis: Methods and Protocols*, Methods in Molecular Biology, vol. 1106, DOI 10.1007/978-1-62703-736-5_11,

and other staphylococci were not widely believed to undergo natural transformation. Isolated reports of transformation in *S. aureus* have been published in the last six decades [2–7], but the mechanism of DNA uptake was never established. Recently, the alternative sigma factor σ^H has been shown to mediate the expression of the *com* genes and transformation in *S. aureus*, although the conditions under which σ^H is active are yet undetermined [7]. Although *S. epidermidis* genomes also harbor a *sigH* ortholog, natural transformation has not yet been described in this species.

Within staphylococcal studies, introduction of plasmid DNA by transformation is not typically performed by induction of the *com* apparatus, but instead by manipulation of the cell membrane through electroporation. Protoplast transformation, in which the cell wall is removed from the recipient cell by lysostaphin digestion, is less frequently used in staphylococci as it is time consuming and inconsistent [8]. Protoplast transformation is also ineffective in *S. epidermidis* due to its resistance to lysostaphin digestion [9]. Electroporation is the process of producing transiently permeabilized cells following exposure to a high voltage electrical field [9]. The electrical field induces the formation of transient pores in the cell membrane through which DNA is transported. The method described below was developed by Augustin and Gotz and yields up to 3×10^5 transformants per μg plasmid DNA [9].

A number of barriers to transformation in staphylococci have been described, including types I and IV restriction–modification systems [10–13] and the production of secreted nucleases that may degrade extracellular DNA. Clustered, regularly interspaced, short palindromic repeat (CRISPR) loci also have been implicated in limiting horizontal gene transfer (particularly conjugation) in *S. epidermidis* [14]. Because of these restriction barriers, to enable the passage of DNA from *E. coli* into the desired staphylococcal strain, *S. aureus* RN4220 must be used as an intermediate host for plasmid DNA first constructed or modified in *E. coli*. RN4220 contains a mutant *hsdR* gene, which encodes part of the Type I restriction endonuclease limiting the uptake of foreign DNA [10]. The *hsdM* and *hsdS* genes, which encode enzymes that catalyze modification of the DNA, are intact in RN4220 [10]. The pattern of methylation protects the DNA from restriction and allows closely related staphylococcal species to recognize the DNA for successful transfer. Therefore, plasmid DNA may be isolated from *S. aureus* RN4220 and transferred by electroporation into the desired *S. epidermidis* strain.

For unknown biological reasons, genetic manipulation in *S. epidermidis* and other coagulase-negative staphylococci is more difficult than *S. aureus*; electroporation frequencies are typically an order of magnitude lower than with *S. aureus*. Therefore, while electroporation is the preferred method of transferring plasmids into *S. epidermidis*, conjugative mobilization is an alternative

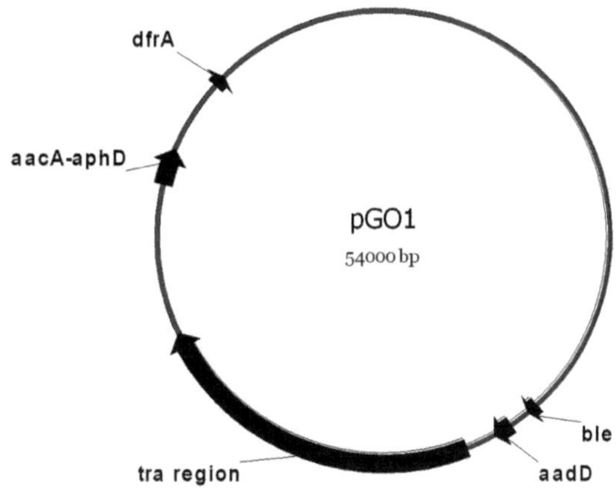

Fig. 1 Plasmid map of pGO1. pGO1 is a 54,000 bp conjugative plasmid required for transfer of DNA from *S. aureus* to *S. epidermidis*. The conjugative transfer gene complex (*tra*) consists of 14 genes and is required for conjugation. The plasmid map also indicates genes that are involved in resistant mechanisms. Bleomycin, kanamycin, gentamicin, and trimethoprim resistances are conferred by the genes *ble*, *aadD*, *aacA-aphD*, and *dfrA*, respectively

well-developed method to introduce plasmid DNA into *S. epidermidis*. Most plasmids used for genetic manipulation in staphylococci do not have conjugative functions, and therefore conjugative plasmids must be introduced into the donor strain (typically *S. aureus* RN4220) to mobilize plasmids by this method.

Conjugative plasmids have long been associated with gentamicin-resistant staphylococcal outbreaks in hospitals, commonly mediating resistance among *S. aureus* and *S. epidermidis* [15–17]. Staphylococcal conjugative plasmids are 40–60 kb in size, encode their own transfer, and typically have a conjugation frequency of 10^{-4}–10^{-5}. The conjugative plasmid pGO1 [18, 19] is one of best-studied staphylococcal conjugative plasmids and has been used to mobilize other plasmids of interest into *S. epidermidis* [20]. Within pGO1, all of the genes required for successful conjugative transfer are contained within a 14.5 kb fragment flanked by *Bgl*II sites and are referred to as the *tra* region (*see* Fig. 1) [19]. Conjugative mobilization of plasmid DNA requires a conjugative plasmid, such as pGO1, to provide the transfer apparatus, and an *oriT* site on the plasmid being mobilized. The system most widely used within the staphylococci was designed by Projan and Archer [20] in 1989 and utilizes pGO1, pC221 (*see* Fig. 2), which provides MobA in trans, and pROJ6448$_{(ts)}$ which contains the pC221 *oriT* site [20]. pC221 is a 4.6 kb plasmid mediating chloramphenicol resistance and contains the *cis*-acting *oriT* and the mobilization genes *mobABC* [20]. The DNA relaxase, MobA, forms relaxation complexes by single-strand nicks within a conserved sequence in *oriT* [21, 22]. The origin of transfer, *oriT*, is located on a fragment

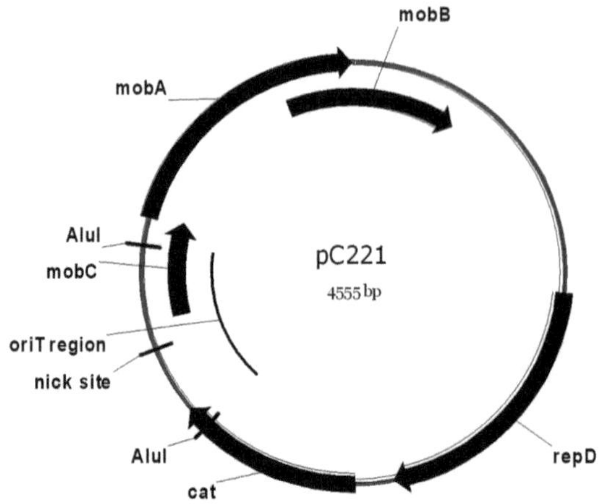

Fig. 2 Plasmid map of pC221. pC221 is a 4,555 bp mobilization plasmid required for the mobilization of pC221 and pROJ6448. This small plasmid contains genes that confer chloramphenicol resistance (*cat*) and genes required in forming the relaxation complexes (*mobCAB*). The origin of transfer (*oriT*) of pC221 is found within a 693 bp *AluI* fragment (noted in figure)

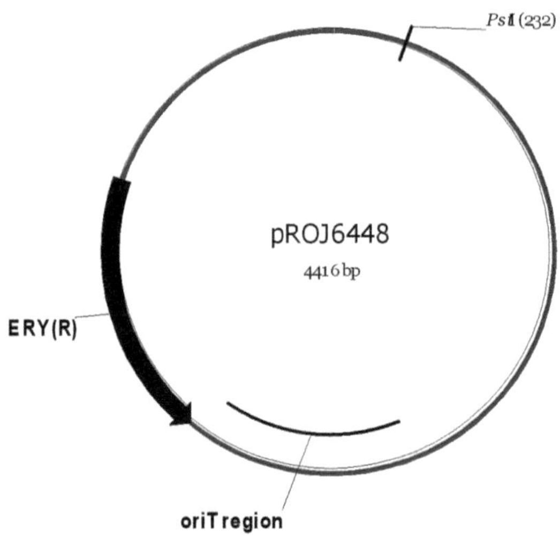

Fig. 3 Plasmid map of $pROJ6448_{(ts)}$. $pE194_{(ts)}$-derived pROJ6448 was constructed by cloning the 693 bp *AluI* fragment from pC221 (containing the *oriT* site) into the unique *ClaI* site of $pE194_{(ts)}$ [21]. Note the *AluI* fragment was blunt-end cloned, thus both the *ClaI* and *AluI* sites are ablated. Also note the unique *PstI* site where pUC19 containing (or other) gram-negative plasmids can be ligated with $pROJ6448_{(ts)}$

flanked by *Alu*I restriction sites and has been transferred to $pE194_{(ts)}$ to create $pROJ6448_{(ts)}$ (*see* Fig. 3) [20]. Thus, when using conjugative mobilization for genetic manipulation, pC221 and pGO1 act in concert to mobilize $pROJ6448_{(ts)}$. Typically, genes of interest are first cloned in *E. coli* using pUC19-based vectors;

pROJ6448$_{(ts)}$ is next ligated into the resulting pUC19 based vectors using the unique *Pst*I site in pROJ6448 thus providing a staphylococcal origin of replication for subsequent transfer into RN4220 and *S. epidermidis*. Additionally, conjugative mobilization may be a useful tool in genetic manipulation of other staphylococcal species where no other genetic transfer system is available.

In this chapter, we provide protocols for transferring staphylococcal plasmids from *S. aureus* to *S. epidermidis* by two methods. Electroporation is the more widely used protocol; however, this methodology is somewhat strain dependent and thus may not be successful in all strain backgrounds. However, our experience suggests that although conjugative mobilization is a more time-consuming experimental process, this method can be used to transfer plasmid DNA into virtually any *S. epidermidis* strain background.

2 Materials

Prepare all solutions using distilled water.

2.1 Preparation of Electrocompetent Cells

1. B (basic) medium: 1 % peptone, 0.5 % yeast extract, 0.5 % NaCl, 0.1 % K_2HPO_4, 0.1 % glucose, pH 7.4. Sterilize by autoclaving, and store at room temperature (*see* **Note 1**).
2. 10 % glycerol: approximately 80 mL per flask of competent cells is required. Sterilize by autoclaving and store at 4 °C.

2.2 Transformation by Electroporation

1. SMMP medium: 75 % SMM medium (1 M sucrose, 40 mM maleic acid, 40 mM $MgCl_2$, pH 6.5, autoclaved for 10 min), 20 % of 7 % Penassay broth (Difco Antibiotic Medium 3, sterilized), 2.5 % of 10 % bovine serum albumin (filtered).
2. Electroporation system.
3. Electroporation cuvettes, presterilized, 1–2 mm gap.

2.3 Conjugative Mobilization

1. Tryptic soy agar (TSA).
2. Sterile 0.9 % NaCl.
3. Sterile 5 mL syringe.
4. Nitrocellulose membrane filters (0.45 μm pore size, 13 mm diameter).
5. Swinnex® filter holder, 13 mm (EMD Millipore Corporation, Billerica, MA, USA). Sterilize by autoclaving.
6. Sterile cotton swabs.
7. Spectrophotometer.
8. Tryptic soy agar (TSA; 4 plates each) supplemented with:
   (a) Novobiocin (1 μg/mL) and rifampin (10 μg/mL).
   (b) Gentamicin (5 μg/mL).

(c) Novobiocin (1 μg/mL), rifampin (10 μg/mL), and gentamicin (5 μg/mL).
(d) Novobiocin (1 μg/mL), rifampin (10 μg/mL), and chloramphenicol (20 μg/mL).
(e) Novobiocin (1 μg/mL), rifampin (10 μg/mL), and erythromycin (10 μg/mL).

3 Methods

3.1 Preparation of Electrocompetent *S. epidermidis*

1. Grow an overnight culture of *S. epidermidis* recipient strain aerobically in 25 mL of B medium (*see* **Note 2**).
2. Dilute the overnight culture to $OD_{600}=0.1$ in 25 mL of fresh, pre-warmed B medium. Grow the cells to $OD_{600}\sim2.0$ at 37 °C with shaking at 250 RPM (*see* **Note 3**).
3. Again, dilute the culture into 50 mL fresh, pre-warmed B medium to $OD_{600}=0.1$.
4. When the cells reach early exponential phase ($OD_{600}=0.5$–0.65), harvest the cells by centrifugation at 5,900 × *g* for 10 min. Centrifugation should be performed at 4 °C.
5. Wash the resulting pellet as follows. First, the cells are resuspended in one volume cold (4 °C) 10 % glycerol. They are harvested again as before; then, the washing process is repeated with ½, 1/20, and 1/50 volumes of cold 10 % glycerol.
6. The final pellet is suspended in 700 μl of cold 10 % glycerol, and the competent cells are divided into 60 μl aliquots. They may be used immediately or stored at −80 °C for future use (*see* **Note 4**).

3.2 Electroporation of *S. epidermidis*

1. Donor plasmid DNA should be of high quality and concentration (500 ng/μL) without contaminating proteins or RNA.
2. If the competent recipient cells have been frozen, thaw at room temperature for 5 min.
3. Next, add 300–1,000 ng plasmid DNA (*see* **Note 5**). Incubate the cell/plasmid DNA mixture at room temperature for 30 min.
4. Perform electroporation at room temperature using an electroporator at 2 kV, 25 μF capacitance, and 100 Ω resistance (*see* **Note 6**).
5. Immediately after electroporation, add the cell suspension to a tube of 950 μL SMMP medium.
6. Incubate the bacteria for 4 h at 37 °C with shaking at 250 RPM then plate on selective medium. If a temperature-sensitive plasmid is being used (such as $pE194_{ts}$-derived plasmids), all incubation steps should be performed at 30 °C.

3.3 Selection and Screening of Transformants

1. Transformants will appear on antibiotic selective medium after 24–48 h of incubation, depending on the method of antibiotic selection being used. Transfer colonies using a sterile toothpick to a fresh plate of selective media.
2. Incubate the selected colonies at 37 °C overnight (or at 30 °C for temperature-sensitive plasmids).
3. Screen potential colonies by PCR or plasmid restriction digest to detect plasmid of interest.

3.4 Conjugative Mobilization

1. Grow the donor (*S. aureus* RN4220 containing pGO1, pC221 and pROJ6448$_{(ts)}$-containing plasmid) and recipient strains (*S. epidermidis* strain of interest) overnight on a TSA plate. The recipient *S. epidermidis* strain should be a novobiocin- and rifampin-resistant derivative of wild-type *S. epidermidis* (*see* **Note 7**). pROJ6448$_{(ts)}$-containing strains should be grown at 30 °C.
2. Prepare the cells by adjusting the cell concentration to an OD_{600} = 1.0 in sterile 0.9 % NaCl.
3. Combine the donor and recipient strain in a 1:10 ratio in a 1 mL volume (i.e., 100 μL of donor cells and 900 μL of recipient cells).
4. Assemble the Swinnex® filter holder with the sterile 0.45 μm nitrocellulose filter. With a 5 mL syringe, push the donor/recipient cells through the filter (*see* **Note 8**).
5. Place the filter on a BHI agar plate with the bacteria face up for 18–24 h. Incubate at 30 or 37 °C depending on the temperature sensitivity of the plasmid used. Staphylococci must be in contact on a surface, as conjugation frequency is decreased if transfer is assessed following growth in broth [16].
6. The following day, resuspend the filter in 1 mL sterile 0.9 % NaCl and vortex to remove bacteria from the filter. Dilute the suspension in sterile saline and plate on TSA with the appropriate antibiotic (*see* **Note 9**).
   (a) Plate 10^{-5}–10^{-8} on TSA containing gentamicin for donor CFU.
   (b) Plate 10^{-5}–10^{-8} on TSA containing novobiocin and rifampin for recipient CFU.
   (c) Plate 10^{-1}–10^{-4} on TSA containing novobiocin, rifampin, and gentamicin to assess conjugation frequency (transfer of pGO1).
   (d) Plate 10^{-1}–10^{-4} on TSA containing novobiocin, rifampin, and chloramphenicol to assess mobilization frequency (transfer of pC221).

(e) Plate 10^{-1}–10^{-4} on TSA containing novobiocin, rifampin, and erythromycin to detect mobilization of the pROJ6448-containing plasmid.

(f) Incubate at 30 or 37 °C depending on temperature sensitivity of plasmid.

7. Plasmid analysis should be performed with colonies obtained in 6e to confirm plasmid transfer. Following confirmation, the pROJ6448$_{(ts)}$-containing plasmid should be transferred by phage transduction into a clean wild-type background (non nov^R rif^R) (*see* **Note 10**).

4 Notes

1. B media works well in this procedure because it contains less glucose than richer media such as tryptic soy broth (TSB) or brain-heart infusion (BHI). This is important because some strains of *S. epidermidis*, such as 1457 [23], produce copious amounts of polysaccharide intercellular adhesin (PIA) in the presence of excess glucose, causing clumping and reducing transformation efficiency.
2. For aerobic growth, use a 1:10 volume:flask ratio (i.e., 25 mL of broth in a 250 mL flask) and 250 RPM shaking in a rotary incubator.
3. Alternatively, Subheading 3.1, **step 2** (the first dilution of the overnight culture) may be omitted if the overnight culture is 12 h or less old. The most important consideration at this step is that the starter culture is in exponential phase of growth. An older starter culture may be in the stationary phase of growth, and thus tends to be less metabolically active and to contain more dead cells. If the first dilution is omitted, dilute the overnight culture to $OD_{600} = 0.1$ in 50 mL B medium, then grow to mid-exponential phase ($OD_{600} = 0.5$–0.65) as written.
4. Fresh competent cells have higher transformation efficiency than previously frozen cells.
5. Transformation efficiency can vary widely with the amount of DNA added. To save time, it is often helpful to use a few different amounts of DNA for the same electroporation process (for example, one tube with 500 ng DNA, one with 700 ng, and one with 900 ng).
6. If the cells/DNA preparation arc during electroporation, the DNA preparation may contain too much salt. To dialyze the DNA, place a small volume of the donor DNA suspension on a 0.025 μm nitrocellulose membrane floating on distilled water. Allow the solutions to equilibrate for about 10 min; then, pipet the DNA suspension off of the membrane. If the problem persists, a new preparation of donor DNA may be necessary.

7. For our studies, we used *S. epidermidis* 1457 as our wild-type strain. To generate the 1457 novobiocin- and rifampin-resistant strain for use as a recipient in the conjugation/mobilization studies, 10^{10} CFU of 1457 was streaked onto a TSA plate containing 10 μg/mL rifampin. Single colonies were picked after 48 h and confirmed to be rifampin resistant by growing again on TSA containing 10 μg/mL rifampin. To generate rifampin- and novobiocin-resistant 1457, 10^{10} cells of the rifampin-resistant 1457 derivative were plated on TSA containing 1 μg/mL of novobiocin, incubated for 48 h, and single rifampin- and novobiocin-resistant colonies were picked and confirmed. Rifampin and novobiocin resistances are easily generated through point mutations in the RNA polymerase β-subunit (*rpoB*) and gyrase B subunit (*gyrB*) genes, respectively [24, 25].
8. If the filter becomes clogged with cells, disassemble the device, remove the filter, and continue the protocol. Alternatively, assemble the Swinnex® device as before and continue the protocol with multiple filters to increase the probability of DNA transfer into *S. epidermidis*.
9. The donor and recipient should have approximately the same number of viable cells (~10^9 CFU). If the donor has significantly more viable cells (1 $\log_{10}$) than recipient, the mobilization frequency of pROJ6448 containing plasmids is virtually undetectable. Therefore, determining the viable count of both donor and recipient CFU (Subheading 3.4, **steps 6a** and **6b**) is an important control until the experimenter has significant experience with specific recipient *S. epidermidis* strain backgrounds. *S. aureus* outcompetes *S. epidermidis*, so the original 1:10 donor to recipient ratio should be closer to a 1:1 ratio when plated. Transfer of pGO1 occurs at a frequency of 10^{-6}–10^{-7}, whereas transfer of pC221 and pROJ6448-containing plasmids is 10^{-7} and 10^{-8}, respectively.
10. We perform allelic replacement experiments in the 1457 nov^R rif^R background before transducing into the 1457 wild-type background to eliminate an unnecessary transduction before recombination. However, the plasmid of interest may be transduced into the 1457 wild-type background before homologous recombination if the mutation being transferred is unmarked.

References

1. Griffith F (1928) The significance of pneumococcal types. J Hyg (Lond) 27:113–159
2. Lindberg M, Sjostrom J-E, Johansson T (1972) Transformation of chromosomal and plasmid characters in *Staphylococcus aureus*. J Bacteriol 109:844–847
3. Nomura H, Udou T, Yoshida K et al (1971) Induction of hemolysin synthesis by transformation in *Staphylococcus aureus*. J Bacteriol 105:673–675
4. Thompson NE, Pattee PA (1977) Transformation in *Staphylococcus aureus*: role of

bacteriophage and incidence of competence among strains. J Bacteriol 129:778–788
5. Thompson NE, Pattee PA (1981) Genetic transformation in *Staphylococcus aureus*: demonstration of a competence-conferring factor of bacteriophage origin in bacteriophage 80α lysates. J Bacteriol 148:294–300
6. Birmingham VA, Pattee PA (1981) Genetic transformation in *Staphylococcus aureus*: isolation and characterization of a competence-conferring factor from bacteriophage 80α lysates. J Bacteriol 148:301–307
7. Morikawa K, Takemura AJ, Inose Y et al (2012) Expression of a cryptic secondary sigma factor gene unveils natural competence for DNA transformation in *Staphylococcus aureus*. PLoS Pathog 8:1–20
8. Gotz F, Schumacher B (1987) Improvements of protoplast transformation in *Staphylococcus carnosus*. FEMS Microbiol Lett 40:285–288
9. Augustin J, Gotz F (1990) Transformation of *Staphylococcus epidermidis* and other staphylococcal species with plasmid DNA by electroporation. FEMS Microbiol Lett 66:203–207
10. Waldron DE, Lindsay JA (2006) Sau1: a novel lineage-specific type I restriction-modification system that blocks horizontal gene transfer into *Staphylococcus aureus* and between *S. aureus* isolates of different lineages. J Bacteriol 188: 5578–5585
11. Corvaglia AR, Francois P, Hernandez D et al (2010) A type III-like restriction endonuclease functions as a major barrier to horizontal gene transfer in clinical *Staphylococcus aureus* strains. Proc Natl Acad Sci U S A 107:11954–11958
12. Xu SY, Corvaglia AR, Chan SH et al (2011) A type IV modification-dependent restriction enzyme SauUSI from *Staphylococcus aureus* subsp. *aureus* USA300. Nucleic Acids Res 39: 5597–5610
13. Monk IR, Shah IM, Xu M et al (2012) Transforming the untransformable: application of direct transformation to manipulate genetically *Staphylococcus aureus* and *Staphylococcus epidermidis*. MBio 3:1–11
14. Marraffini LA, Sontheimer EJ (2008) CRISPR interference limits horizontal gene transfer in staphylococci by targeting DNA. Science 322: 1843–1845
15. Archer GL, Johnston JL (1983) Self-transmissable plasmids in staphylococci that encode resistance to aminoglycosides. Antimicrob Agents Chemother 24:70–77
16. Forbes BA, Schaberg DR (1983) Transfer of resistance plasmids from *Staphylococcus epidermidis* to *Staphylococcus aureus*: evidence for conjugative exchange of resistance. J Bacteriol 153:627–634
17. Jaffe HW, Sweeney HM, Nathan C et al (1980) Identity and interspecific transfer of gentamicin-resistance plasmids in *Staphylococcus aureus* and *Staphylococcus epidermidis*. J Infect Dis 141: 738–747
18. Caryl JA, O'Neill AJ (2009) Complete nucleotide sequence of pGO1, the prototype conjugative plasmid from the Staphylococci. Plasmid 62:35–38
19. Thomas WD, Archer G (1989) Identification and cloning of the conjugative transfer region of *Staphylococcus aureus* plasmid pGO1. J Bacteriol 171:684–691
20. Projan SJ, Archer GL (1989) Mobilization of the relaxable *Staphylococcus aureus* plasmid pC221 by the conjugative plasmid pGO1 involves three pC22 loci. J Bacteriol 171:1941–1845
21. Novick R (1976) Plasmid-protein relaxation complexes in *Staphylococcus aureus*. J Bacteriol 127:1177–1187
22. Caryl JA, Smith MCA, Thomas CD (2004) Reconstitution of a staphylococcal plasmid-protein relaxation complex in vitro. J Bacteriol 186:3374–3383
23. Mack D, Fischer W, Krokotsch A et al (1996) The intercellular adhesin involved in biofilm accumulation of *Staphylococcus epidermidis* is a linear beta-1,6-linked glucosaminoglycan: purification and structural analysis. J Bacteriol 178:175–183
24. Aubry-Damon H, Soussy C-J, Courvalin P (1998) Characterization of mutations in the *rpoB* gene that confer rifampin resistance in *Staphylococcus aureus*. Antimicrob Agents Chemother 42:2590–2594
25. Stieger M, Angehrn P, Wohlgensinger B et al (1996) GyrB mutations in *Staphylococcus aureus* strains resistant to cyclothialidine, coumermycin, and novobiocin. Antimicrob Agents Chemother 40:1060–1062

Chapter 12

Methods to Generate a Sequence-Defined Transposon Mutant Library in *Staphylococcus epidermidis* Strain 1457

Todd J. Widhelm, Vijay Kumar Yajjala, Jennifer L. Endres, Paul D. Fey, and Kenneth W. Bayles

Abstract

Transposon mutant libraries are valuable resources to investigators studying bacterial species, including *Staphylococcus epidermidis*, which are difficult to genetically manipulate. Although sequence-defined transposon mutant libraries have been constructed in *Staphylococcus aureus*, no such library exists for *S. epidermidis*. Nevertheless, the study of Tn*917*-mediated mutations has been paramount in discovering unique aspects of *S. epidermidis* biology including initial adherence and accumulation during biofilm formation. Herein, we describe modifications to the methodology first described by Bae et al. to utilize the *mariner*-based transposon *bursa aurealis* to generate mutants in *S. epidermidis* strain 1457.

Key words *Bursa aurealis*, Transposon, *Staphylococcus epidermidis*

1 Introduction

It is well known in the staphylococcal community that generating allelic replacement mutations in *Staphylococcus epidermidis* is especially difficult when compared to performing the same procedures in *S. aureus*. Specifically, *S. epidermidis* has a low electroporation efficiency compared to *S. aureus* and detection of allelic replacement using temperature-sensitive plasmids is extremely inefficient. Given these two major obstacles, generation of mutations can be an exceptionally long and cumbersome process from the beginning of the project to the final confirmation of a proper mutation. The construction of sequence-defined transposon mutant libraries is one strategy to overcome these obstacles.

The *Enterococcus faecalis* transposon Tn*917* has been used extensively to discover loci that function in initial adherence to surfaces and biofilm accumulation in *S. epidermidis* [1–3]. Indeed, Knobloch and colleagues developed an arbitrary primer PCR approach to further enhance rapid identification of loci interrupted by Tn*917* [4]. However, Bae and colleagues determined that

Paul D. Fey (ed.), *Staphylococcus Epidermidis: Methods and Protocols*, Methods in Molecular Biology, vol. 1106,
DOI 10.1007/978-1-62703-736-5_12, © Springer Science+Business Media, LLC 2014

Tn*917* exhibited significant bias into two regions of the *S. aureus* chromosome; this same bias was not observed using the mariner-derived transposon *bursa aurealis* [5]. *Bursa aurealis* has recently been utilized to construct two sequence-defined libraries in *S. aureus*; one in strain Newman [5, 6] and the other in strain USA300 JE2 [7, 8]. The use of both of these libraries has been extremely helpful in the screening of phenotypes, virulence in models of infection, or rapid dissection of metabolic pathways [5, 8–10]. Furthermore, a detailed experimental methodology to generate *bursa aurealis* mutations in *S. aureus* has been outlined [6].

To date, a completely accessible library is not available to investigators studying *S. epidermidis*, particularly due to the difficult nature of performing any genetic manipulations in this species. However, during the optimization and construction of the Nebraska Transposon Mutant Library in *S. aureus* JE2 [8], conditions necessary for generating *bursa aurealis* mutants in *S. epidermidis* 1457 [11] were tested and successfully generated. Following is a detailed protocol to generate *bursa aurealis* insertion mutants in *S. epidermidis* 1457 and identify their transposon–genome junction site.

2 Materials

Prepare all reagents using ultrapure water and analytical grade reagents. Prepare and store all reagents at room temperature (unless otherwise indicated).

2.1 Components Required for Heat Shock and Patching

1. Ultrapure water.
2. Tryptic soy agar (TSA).
3. Incubator and heat block set at 46.5 °C.
4. Antibiotic stocks: Chloramphenicol (10 mg/ml), erythromycin (2.5 mg/ml), and tetracycline (5 mg/ml).
5. 1.5 ml microcentrifuge tubes.

2.2 Components for Genomic DNA Isolation

1. Tryptic soy broth (TSB).
2. Erythromycin (2.5 mg/ml).
3. Shaking incubator set at 37 °C.
4. Fifty percent glycerol.
5. Colony Picker (V & P Scientific, Inc.) or toothpicks.
6. 1 ml 96-deep-well polypropylene plate (Thermo Scientific).
7. 2 ml 96-deep-well plate (Nunc).
8. Wizard Genomic DNA purification Kit (Promega, Madison, WI).
9. 50 mM EDTA.

10. Lysostaphin (AMBI Products LLC, Lawrence, NY).
11. Tris–EDTA Buffer—[For 100 ml; 1 ml of 1 M Tris–HCl (pH 8.0) + 0.5 ml of 0.5 M EDTA (pH 8.0) + 98.5 ml water].
12. 70 % ethyl alcohol.
13. Isopropyl alcohol (isopropanol).

2.3 Molecular Genetic Components to Confirm Transposon Insertion Site

1. 96-well PCR plates.
2. Semi-skirted 96-well PCR plates.
3. AciI restriction enzyme plus appropriate enzyme buffer.
4. Ligation master mix: 2.5 μl 10× T4 ligase buffer (Monserate Biotechnology Group), 0.5 μl Dilution buffer (Monserate Biotechnology Group), 0.5 μl T4 ligase (Monserate Biotechnology Group), 1.5 μl nuclease-free water.
5. Taq polymerase (Monserate Biotechnology Group, San Diego, CA).
6. Forward primer (Buster) GCTTTTTCTAAATGTTTTTTAA GTAAATCAAGTACC [6].
7. Reverse primer (Martin ermR) AAACTGATTTTTAGTAAA CAGTTGACGATATTC [6].
8. Thermocycler.
9. ExoSAP-IT (GE healthcare Life Sciences).
10. 1 % agarose gel.

3 Methods

S. epidermidis 1457 carrying pFA545 (encodes the *bursa aurealis* transposase; [5, 6]) is transduced with a phage ϕ71 [12] that has been propagated on 1457 carrying pBursa (encoding the *bursa aurealis* [5, 6]). Following incubation at 30 °C (due to temperature sensitivity of both pBursa and pFA545), individual chloramphenicol (encoded by pBursa) and tetracycline (encoded by pFA545) resistant colonies are then subsequently heat shocked using the following protocol (*see* **Note 1**).

3.1 Heat Shock to Cure Plasmids pBursa and pFA545 and Detect Transposition Event

1. Aliquot 1 ml of sterile ultrapure water into ten 1.5 ml microcentrifuge tubes.
2. Place tubes in a heat block at 46.5 °C for at least 1 h.
3. With a sterile cotton swab, pick 1–2 transductants per tube and gently resuspend the colonies in the water to an OD_{600} value between 2 and 4 (*see* **Note 2**).
4. Plate 200 μl of the sample onto pre-warmed (46.5 °C) TSA plates containing 2.5 μg/ml erythromycin and incubate at 46.5 °C for 2 days (*see* **Note 3**).

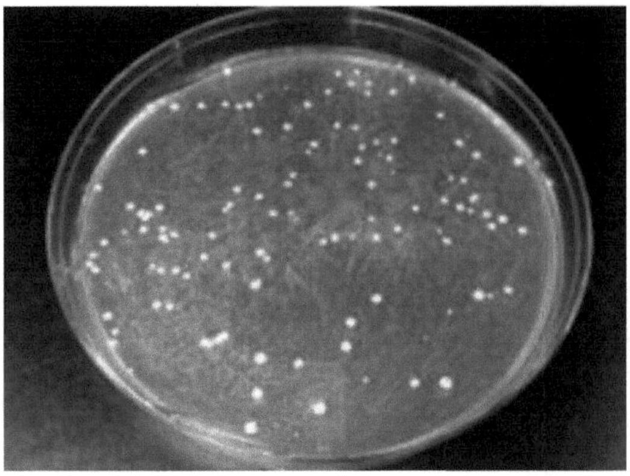

Fig. 1 TSA plate containing 2.5 μg/ml erythromycin following heat shock and incubation at 46.5 °C. Note both large colonies and background (small colonies) are present

3.2 Patching for Agar-Based Selection of Mutants

1. Following 2 days incubation at 46.5 °C, pick colonies onto a TSA plate containing 2.5 μg/ml erythromycin (Fig. 1).
2. Patch the colonies obtained from heat shock experiment onto three TSA plates each with a different selection marker (erythromycin 2.5 μg/ml, chloramphenicol 10 μg/ml and tetracycline 5 μg/ml) and incubate at 46.5 °C.
3. Pick only those colonies that are resistant to erythromycin but susceptible to chloramphenicol and tetracycline. These results suggest that the isolate has lost both pBursa and pFA545 but *bursa aurealis* (encoding erythromycin resistance) has inserted randomly into the chromosome (*see* **Note 4**).

3.3 Identification of Transposon Insertion Sites by Inverse PCR and DNA Sequencing

3.3.1 Isolation of Genomic DNA by Modified Promega Wizard Genomic Prep Protocol

Day 1:

1. Fill the 1 ml, 96-well plate with 400 μl TSB with erythromycin (2.5 μg/ml).
2. Using the colony picker or toothpicks, inoculate the probable *bursa aurealis* mutants (erythromycin resistant, chloramphenicol and tetracycline susceptible) into the wells containing TSB with erythromycin (2.5 μg/ml) and shake vigorously (250 rpm) at 37 °C overnight.

Day 2:

1. Centrifuge the 96-well plate to pellet the cells (3,000 × *g* for 10 min).
2. Discard supernatant and resuspend pellets in 110 μl of 50 mM EDTA.
3. Add 10 μl of a 10-mg/ml solution of lysostaphin and mix vigorously.

4. Incubate at 37 °C for 90 min (mixture should become gelatinous and translucent).
5. Add 600 μl Promega Nuclei Lysis Buffer and incubate the plate at 80 °C for 10 min (*see* **Note 3**).
6. Cool to room temperature and then add 200 μl of Promega Protein precipitation solution. Vortex vigorously for 2 min and then place on ice for 10 min.
7. Centrifuge at 3,000 × *g* for 10 min.
8. Transfer supernatant (without disturbing pellet) into a 2 ml 96-well plate containing 600 μl ice-cold isopropanol. Mix well until all components are evenly distributed.
9. Centrifuge at 3,000 × *g* for 10 min to collect precipitated DNA.
10. Discard supernatant, being careful not to disrupt the DNA pellet. Add 600 μl ice-cold 70 % ethanol and invert 5–10 times to wash the pellet. Centrifuge at 3,000 × *g* for 10 min.
11. Discard ethanol wash and dry the pellet by leaving the plate open for 15–20 min until all ethanol is evaporated.
12. Dissolve the DNA pellet in 100 μl TE buffer and incubate the plate at 65 °C for 1 h.
13. Store genomic DNA at −20 °C until further use.

Day 3:

AciI Digest of Genomic DNA and Ligation

1. Digests are performed in 20 μl total volume in a 96-well PCR plate. Add 17 μl genomic DNA, 2 μl 10× AciI restriction enzyme buffer, and 1 μl AciI restriction enzyme.
2. Incubate at 37 °C for 2 h and subsequently heat inactivate AciI at 65 °C for 20 min.
3. Aliquot 5 μl of ligation master mix into each reaction and mix (total volume 25 μl)
4. Incubate overnight at 4 °C.

Day 4:

Inverse PCR Protocol and Purification of Amplified DNA

An inverse polymerase chain reaction is used for the identification of the *bursa aurealis* insertion site within the genome using primers that anneal to two different regions on the transposon. After AciI digestion and ligation with T4 ligase, the genome is a collection of circular DNA fragments of various sizes. Some of these circular fragments will contain *bursa aurealis*, thereby supplying a known sequence, allowing primers to bind and amplify the entire circumference of the circular DNA molecule. PCR fragments are then sequenced to identify the *bursa aurealis*–genome junction site.

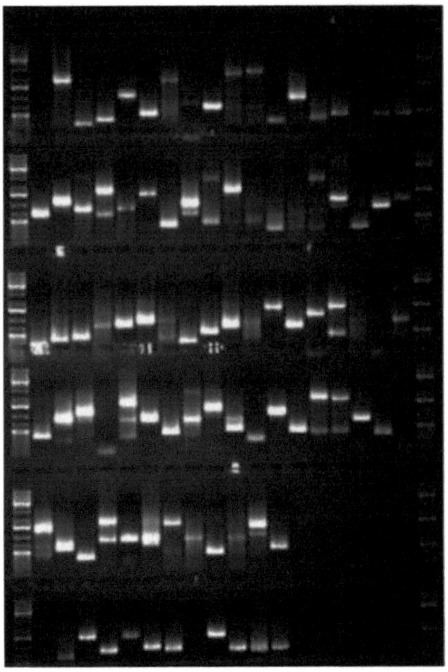

Fig. 2 Image of an 1 % agarose gel loaded with PCR products following *bursa aurealis* mutagenesis protocol. Note the random size of the PCR fragments demonstrating random insertion of *bursa aurealis* into the *S. epidermidis* chromosome

1. Add the following components together for the PCR reaction to amplify the *bursa aurealis*–chromosome junction site. 5 μl DNA digested with AciI and ligated with T4 DNA ligase, 5 μl 10× Taq polymerase buffer, 1 μl Taq polymerase, 1 μl 10 μM forward primer (Buster), 1 μl 10 μM reverse primer (martin ermR), 1 μl dNTPs, 1 μl 50 mM $MgCl_2$, 35 μl nuclease-free water.
2. Perform the amplification using 40 cycles at 94 °C for 30 s, 63 °C for 30 s, and 72 °C for 3 min.
3. 10 μl of subsequent PCR reaction is analyzed on a 1 % agarose gel. If multiple *bursa aurealis*–chromosome junctions are being assessed at the same time, amplicon sizes should be variable due to randomness of *bursa aurealis* insertion (Fig. 2).

Purification of Inverse PCR Products for Sequencing

1. Once confirmation of random insertions through the banding pattern on the gel is obtained, aliquot 6 μl of the PCR product into the 96-well sequencing plates (semi-skirted). Add 2 μl of EXOSAP-IT, to each sample and incubate at 37 °C for 15 min and then 80 °C for 15 min.
2. Use the buster primer to obtain DNA sequence data of the *bursa aurealis*–genome junction site.
3. After high quality DNA sequence files are obtained in chromatogram format, use FinchTV (A brilliant trace viewer from

Geospiza, Inc.) to identify the insertion site of the transposon by BLAST (Bae et al. [6])

(i) To start the query of a particular sequence, manually look for a substring sequence of CCTGTTA which marks the end of the *bursa aurealis* transposon.

(ii) Select 100–200 nucleotides downstream of the Transposon end site CCTGTTA and subject it to a Nucleotide BLAST.

(iii) Upon receiving BLAST results, scroll down to find the reference genome that is most closely related to the strain in which you are conducting transposon mutagenesis and select it; typically for *S. epidermidis* this would be strain RP62A or ATCC12228.

(iv) The following information is obtained, which is tabulated in an excel sheet to calculate the transposon insertion sites and orientation:

(a) Location of the transposon insertion site within the chromosome.

(b) The gene or the intergenic region in which the transposon has inserted.

(c) The orientation of the transposon relative to the gene (if inserted into a gene).

(d) The start and end coordinates of the gene (if inserted into a gene).

(e) The orientation of transposon whether it is on the sense or antisense strand of the genome using the information provided by BLAST as Plus/Plus for sense and Plus/Minus for antisense.

(f) If the gene and the transposon are in the same orientation, then one can utilize the green fluorescent protein (*gfp*) located on *bursa aurealis* as a transcriptional fusion.

4 Notes

1. To guard against transposition of *bursa aurealis* and subsequent competition between unique mutants within a freezer stock of 1457 carrying both pBursa and pFA545, 1457 pBursa and 1457 pFA545 strain stocks are kept separate. Therefore, once pBursa is transduced into 1457 carrying pFA545; heat shock is then performed with individual colonies to isolate unique *bursa aurealis* mutants from each colony. In addition, more transposition events are detected following heat shock when transductant colonies (carrying both pBursa and pFA545) are allowed to sit at 4 °C for 5–7 days.

2. *S. epidermidis* 1457 colonies express high levels of PIA (Polysaccharide Intercellular Adhesin), which makes it difficult for the cells to go into suspension. However, this seems to have no effect on transposition efficiency.
3. Transposition frequency of *bursa aurealis* in *S. epidermidis* 1457 occurs at ~10^{-6}; therefore, 100–200 colonies should be present following incubation at 46.5 °C on TSA containing 2.5 μg/ml erythromycin.
4. In our experience, approximately 90 % of colonies are both chloramphenicol and tetracycline susceptible demonstrating adequate curing of the temperature sensitive plasmids pBursa and pFA545.

References

1. Grueter L, Koenig O, Laufs R (1991) Transposon mutagenesis in *Staphylococcus epidermidis* using the *Enterococcus faecalis* transposon Tn*917*. FEMS Microbiol Lett 66: 215–218
2. Heilmann C, Gerke C, Perdreau-Remington F et al (1996) Characterization of Tn*917* insertion mutants of *Staphylococcus epidermidis* affected in biofilm formation. Infect Immun 64:277–282
3. Mack D, Nedelmann M, Krokotsch A et al (1994) Characterization of transposon mutants of biofilm-producing *Staphylococcus epidermidis* impaired in the accumulative phase of biofilm production: genetic identification of a hexosamine-containing polysaccharide intercellular adhesin. Infect Immun 62:3244–3253
4. Knobloch JK, Nedelmann M, Kiel K et al (2003) Establishment of an arbitrary PCR for rapid identification of Tn*917* insertion sites in *Staphylococcus epidermidis*: characterization of biofilm-negative and nonmucoid mutants. Appl Environ Microbiol 69:5812–5818
5. Bae T, Banger AK, Wallace A et al (2004) *Staphylococcus aureus* virulence genes identified by *bursa aurealis* mutagenesis and nematode killing. Proc Natl Acad Sci U S A 101:12312–12317
6. Bae T, Glass EM, Schneewind O et al (2008) Generating a collection of insertion mutations in the *Staphylococcus aureus* genome using bursa aurealis. Methods Mol Biol 416:103–116
7. Bose JL, Fey PD, Bayles KW (2013) Genetic tools to enhance the study of gene function and regulation in *Staphylococcus aureus*. Appl Environ Microbiol 79:2218–2224
8. Fey PD, Endres JL, Yajjala VK et al (2013) A genetic resource for rapid and comprehensive phenotype screening of nonessential *Staphylococcus aureus* genes. mBio 4:e00537–00512
9. Li C, Sun F, Cho H et al (2010) CcpA mediates proline auxotrophy and is required for *Staphylococcus aureus* pathogenesis. J Bacteriol 192:3883–3892
10. Nuxoll A, Halouska S, Sadykov M et al (2012) CcpA regulates arginine biosynthesis in *Staphylococcus aureus* through repression of proline catabolism. PLoS Pathog 11:e1003033
11. Mack D, Siemssen N, Laufs R (1992) Parallel induction by glucose of adherence and a polysaccharide antigen specific for plastic-adherent *Staphylococcus epidermidis*: evidence for functional relation to intercellular adhesion. Infect Immun 60:2048–2057
12. Nedelmann M, Sabottke A, Laufs R et al (1998) Generalized transduction for genetic linkage analysis and transfer of transposon insertions in different *Staphylococcus epidermidis* strains. Zentralbl Bakteriol 287:85–92

Chapter 13

Examination of *Staphylococcus epidermidis* Biofilms Using Flow-Cell Technology

Derek E. Moormeier and Kenneth W. Bayles

Abstract

A common in vitro method to study *Staphylococcus epidermidis* biofilm development is to allow the bacteria to attach and grow on a solid surface in the presence of a continuous flow of nutrients. Under these conditions, the bacteria progress through a series of developmental steps, ultimately forming a multicellular structure containing differentiated cell populations. The observation of the biofilm at various time-points throughout this process provides a glimpse of the temporal changes that occur. Furthermore, use of metabolic stains and fluorescent reporters provides insight into the physiologic and transcriptional changes that occur within a developing biofilm. Currently, there are multiple systems available to assess biofilm development, each with advantages and disadvantages depending on the questions being asked. In this chapter, we describe the use of two separate flow-cell systems used to evaluate the developmental characteristics of staphylococcal biofilms: the FC270 flow-cell system from BioSurface Technologies, Corp. and the BioFlux1000 microfluidic flow-cell system from Fluxion Bioscience, Inc.

Key words Biofilm, BioFlux, Flow-cell, Epifluorescence microscopy, Confocal laser scanning microscopy

1 Introduction

Given the frequency with which biofilms are associated with bacterial infections in humans and animals [1, 2], the study of biofilm development has become important to understand the mechanisms employed by them to evade the host immune response as well as to survive in the face of antibiotic therapy [3, 4]. One of the most common causes of biofilm-related infections is *Staphylococcus epidermidis*, a harmless member of the skin microbial flora, but an opportunistic pathogen when the epithelial barrier is breached [5]. Since the ability of *S. epidermidis* to form a biofilm is the primary factor required for its virulence, choosing the appropriate biofilm assay to monitor biofilm formation is critical to elucidate the mechanisms used by these bacteria to cause infection.

The gold standard for assessing developmental processes associated with biofilm formation are flow-cell systems that allow for

Paul D. Fey (ed.), *Staphylococcus Epidermidis: Methods and Protocols*, Methods in Molecular Biology, vol. 1106,
DOI 10.1007/978-1-62703-736-5_13,

the perfusion of media across bacterial cells attached to a synthetic surface, providing a constant supply of nutrients under the pressure of a shear force. Today, several commercially available flow-cell systems from companies such as BioSurface Technologies, Corp., Stovall Life Science, Inc., and Fluxion Biosciences, Inc. are available for use in biofilm studies. In fact, a variety of flow-cell systems ranging from single- to multichannel designs are available in either reusable or disposable forms. While some flow-cell systems allow researchers to test the effects of various surfaces such as glass slides/coverslips, polycarbonate coupons, and plastic capillary tubing on biofilm development, others are designed to maximize versatile image acquisition using a high-throughput plate format. In this chapter, we compare and contrast two different flow-cell systems used to examine biofilm formation by *Staphylococcus* sp.: the FC270 flow-cell system from BioSurface Technologies, Corp. and a new automated flow-cell system, BioFlux1000 (BioFlux) from Fluxion Biosciences, Inc.

First, the FC270 flow-cell system is a reusable polycarbonate flow-cell apparatus containing two viewing chambers, each housing two polycarbonate coupons and a glass coverslip. Within the apparatus, bacterial growth media are perfused across a surface on which bacteria are adhered, providing both a large growing surface and a viewing area. The FC270 system is particularly useful for the analysis of resulting biofilms using confocal laser scanning microscopy (CLSM), allowing for the three-dimensional visualization of fluorescent promoter fusion constructs and stains, as well as quantitative data analysis using the COMSTAT software package [6, 7]. In addition, the FC270 system has the capacity to exchange surfaces so that different materials can be tested for their ability to support biofilm growth. However, there are three major drawbacks of the FC270 system: (1) a limited ability to visualize the biofilm during development, (2) only two biofilms can be grown per experiment, and (3) a large volume of growth medium is required.

In contrast to the FC270 system, the BioFlux1000 is comprised of an epifluorescence microscope equipped with an automated temperature-controlled stage, a pneumatic compressor, a high-resolution camera, and specialized 24-well or 48-well plates equipped with microfluidic channels that connect two adjacent wells, one for sterile media and the other for effluent [8]. The microfluidic channels are seeded with bacteria and the sterile media is pneumatically perfused through the channels, resulting in biofilm growth. To assess biofilm development in each channel (up to 8 or 24 biofilms, depending on the plate being used), sequential images are automatically acquired using the high-resolution camera and compiled using BioFlux image analysis software. Together, the automated image acquisition and the simultaneous growth of multiple biofilms enable an unprecedented comparison of biofilm development by bacterial strains containing different mutations [9].

Additionally, the use of metabolic stains and/or fluorescent reporters allows for the localization of spatial and temporal patterns of metabolic activity and gene expression within the developing biofilm architecture [8, 9]. Although the evaluation of the biofilm images is primarily qualitative, the BioFlux data analysis software contains some functions that enable areas of the two-dimensional images of light and/or fluorescence intensity to be quantified [8].

One disadvantage of the BioFlux1000 system is the inability to collect effluent for analysis of metabolic products generated during biofilm growth due to the inability to access the effluent wells during a run. Moreover, although the plates provide a convenient format for high-throughput analysis, the inability to test biofilm maturation on different synthetic surfaces also provides another distinct disadvantage when compared to flow cells with interchangeable surfaces. Additionally, the smallness of the microfluidic channels and media wells can constrain both the growing surface area and media volume used for an individual experiment. More specifically, the channel size (depth, 70 μm; width, 370 μm [8]) can inhibit the size and some architectural morphologies of biofilm development, while media volume constraints limit time of biofilm growth to several hours versus days in the FC270 flow-cell.

Undoubtedly, both the BioSurface Technologies and the BioFlux flow-cell systems provide obvious advantages and disadvantages when studying the intricacies of *S. epidermidis* biofilm. Indeed, studies using both flow-cell systems may provide better insight into biofilm development than using a single system alone. Therefore, as researchers design experimental procedures to test their hypotheses, they must take into account the limitations of each flow-cell system and fine tune each in a way that best addresses their experimental needs.

2 Materials

2.1 BioFlux1000 Flow-Cell Assay

1. Tryptic soy agar (TSA) (EMD).
2. Tryptic soy broth (TSB) (EMD).
3. BioFlux biofilm media: 50 % TSB (v/v) in distilled water.
4. BioFlux 1000 (Fluxion Biosciences Inc.).
5. BioFlux1000 48 Well Plate 0–20 dyn/cm^2 (BioFlux Plate) (P/N 910-0047 Fluxion Biosciences, Inc.).
6. Spectrophotometer.

2.2 FC270 Flow-Cell Assay

1. FC270: BST FC 270 dual channel flat plate flow-cell apparatus (#FC 270, BioSurface Technologies Corp.).
2. Tryptic soy agar (TSA) (EMD Millipore).
3. Biofilm media: 25 % TSB (v/v) in distilled water.

4. Rainin Dynamx RP-1 peristaltic pump.
5. Micro cover glass (coverslip) 24×60 mm (#48382-139, VWR).
6. "Y" manifold (#TFP-BY125-25, Small Parts).
7. Polypropylene straight connector, 1/8″ (#195030000, Bel-Art).
8. Bubble traps (#FC33, BioSurface Technologies Corp.).
9. Ratchet tubing clamps (#EW-06833-00, Cole-Parmer).
10. 5 ml syringe, Luer-Lok Tip (#309646, BD).
11. Incubator or temperature-controlled room set at 37 °C.
12. PVC flow-measured pump tubing 3.18 mm inner-diameter (ID) (#116-0549-21, Pulse Instrumentation).
13. Precision pump tubing 1/8 in. ID (#06434-16, MasterFlex).
14. 4 l Nalgene carboy (Stovall Life Science, Inc.) (#15-453-237, Fisher).
15. Autoclavable syringe filter, 0.2 μm (#28145-477, VWR).
16. 3 ml syringe Luer-Lok Tip with BD PrecisionGlide Needle 25G×1 (3 ml syringe with needle) (#309581, BD).
17. Silicone glue.
18. Aluminum foil or autoclavable paper.
19. Polycarbonate coupons (#RD122-PC, BioSurface Technologies Corp.).

3 Methods

3.1 BioFlux1000 Flow-Cell Assay

For simplicity and ease of explanation, only a protocol for a 48-well BioFlux plate setup is described.

1. Streak bacteria for single colonies on TSA plate and grow overnight at 37 °C.
2. From single colonies, grow overnight cultures in 3 ml of TSB medium with shaking (250 rpm) at 37 °C.

3.1.1 48-Well BioFlux Plate Preparation

1. Prepare inoculums by adjusting overnight cultures to an optical density at 600 nm (OD_{600}) of 0.8 in TSB (*see* **Note 1**).
2. Pipette 200 μl of TSB into output wells of BioFlux plate, being careful not to allow bubbles when pipetting (Fig. 1) (*see* **Note 2**). Place BioFlux plate onto automated-heated stage of microscope. With gasket attached to bottom BioFlux Interface, carefully clamp down the BioFlux Interface to the plate (Fig. 2) (*see* **Note 3**).
3. Prime desired channels in BioFlux plate by pumping 200 μl TSB from the output well to the input well for 5 min at

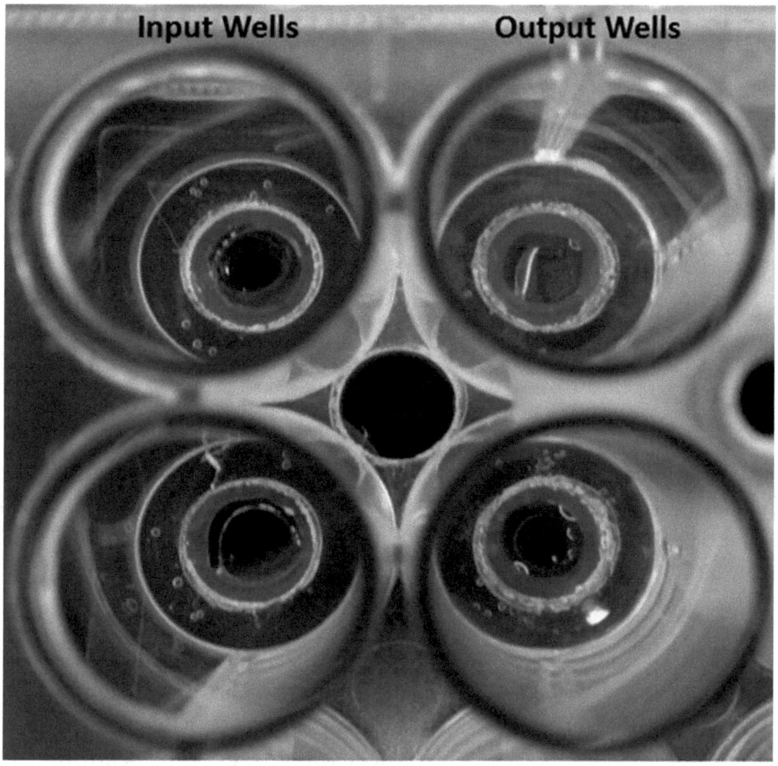

Fig. 1 Macroscopic view of input and output wells of a 48-well BioFlux plate with media in the center. *Circle* in the center of wells is the viewing window of the microfluidic channels. To prime channels, 200 μl of TSB is pumped for 5 min from output to input wells. After excess TSB is aspirated from output, 200 μl of adjusted inoculums are seeded into the channels by pumping for 5 s from output to input wells. After 1 h incubation, biofilms are grown in 50 % TSB pumped from input to output wells for 18 h at 0.6 dyn/cm^2 (64 μl/h)

5.00 dyn/cm^2. To ensure the pump in the BioFlux Controller is working correctly, green lights on the front of the BioFlux Controller will indicate the exact port that is pumping (Continuous running of compressor indicates a leak in the tubing or interface. If this occurs, stop pumping and fix the leak before continuing.).

4. Stop priming after 5 min and unclamp the BioFlux interface. Ensure the TSB has successfully been pumped through the channels and can be seen in the center of the input wells in the BioFlux plate. Aspirate off excess TSB from the output wells being careful not to disturb the TSB in the center of the well.
5. Seed the channels by adding 200 μl of adjusted OD_{600} inoculums into the output wells and 300 μl of 50 % TSB into the input wells. Place the BioFlux plate back on the stage, and

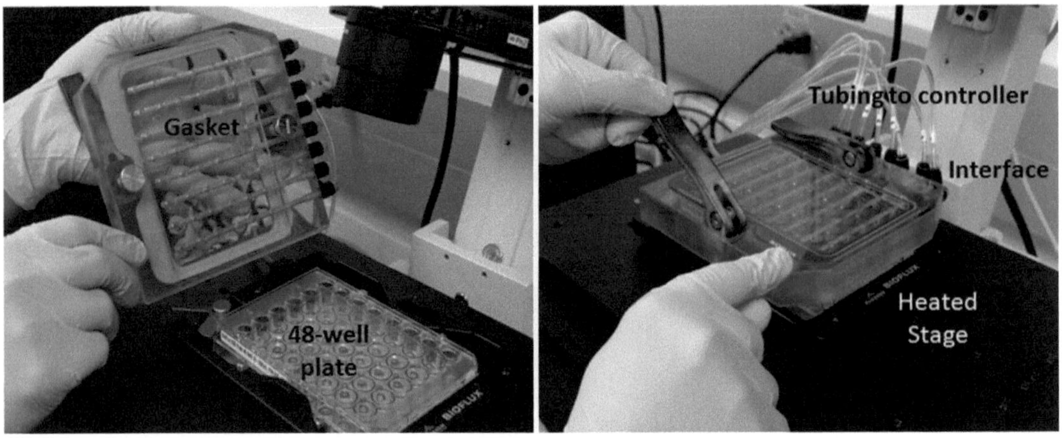

Fig. 2 The BioFlux Interface locks the BioFlux plate to the heated stage and is connected to the BioFlux Controller via small tubing. The tubing allows media to be pneumatically pumped into the microfluidic channels of the BioFlux plate. It is important during every attachment of the BioFlux Interface to the BioFlux Plate to apply as little pressure as possible to the top of the Interface to prevent the plate from cracking

clamp down the BioFlux interface. Pump from output to input wells for 4–5 s at 2.00 dyn/cm^2 and ensure that all channels are successfully seeded by scanning the channels of the plate. If the channels are not seeded, pump for another 1–2 s until the bacterial cells cover the channel area. Allow the cells to attach to the bottom of the channels by incubating on the heated stage at 37 °C for 1 h.

6. After 1 h incubation, remove the BioFlux interface and aspirate off the cultures from the output wells, being careful not to disturb the media in the center of wells. Add 1 ml of 50 % TSB to the 300 μl in the input wells, place back on the stage and clamp down the BioFlux Interface. The plate is now ready for image acquisition.

3.1.2 Calibration of BioFlux Plate and Automation Flow Setup

1. Using BioFlux Montage Software, calibrate the BioFlux plate for viewing channels at the desired magnification, using steps described in the BioFlux1000 Operation Manual. To view the channel from top-to-bottom, we suggest imaging using the 20× objective (200× magnification).
2. After calibrating and ensuring that all the viewing areas are in focus, under the Multi-Dimensional Acquisition automation software, select the program you want to use to run the BioFlux1000. For our strains, we design a program that pumps media from the input to output wells for 18 h at 0.6 dyn/cm^2 (64 μl/h) and captures bright field and/or epifluorescent images for every channel in 5-min intervals (*see* **Note 4**).

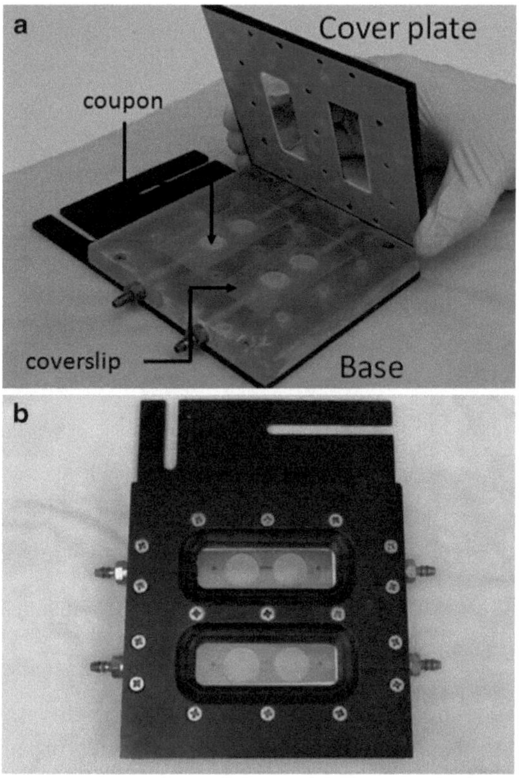

Fig. 3 (**a**) Four polycarbonate coupons are placed into recessed flow channels, followed by carefully placing two glass coverslips into each indentation of the base. With the silicone gasket attached, the cover plate is then gently placed on top of the base with screw holes aligned. 17 screws are then placed into holes and screwed down until almost snug and then autoclaved. (**b**) FC270 flow-cell apparatus shown fully assembled

3.2 FC270 Flow-Cell Assay

1. Streak bacteria for single colonies on TSA plates and grow overnight at 37 °C.
2. From single colonies, grow overnight cultures in 3 ml of TSB with shaking (250 rpm) at 37 °C.

3.2.1 Assembly of the FC270 Flow-Cell Apparatus

1. Insert four polycarbonate coupons into the recessed flow-cell channels and place two glass coverslips over the coupons into each indention of the base (Fig. 3a).
2. Align holes of the silicone gasket with holes on the aluminum cover plate and gently smooth out the gasket until flush. Wetting the gasket briefly with water will allow it to more securely attach to the cover plate (Fig. 3a).
3. Carefully, line up one edge of the cover plate with the base and lower the cover plate onto the base, keeping screw holes aligned (*see* **Note 5**).

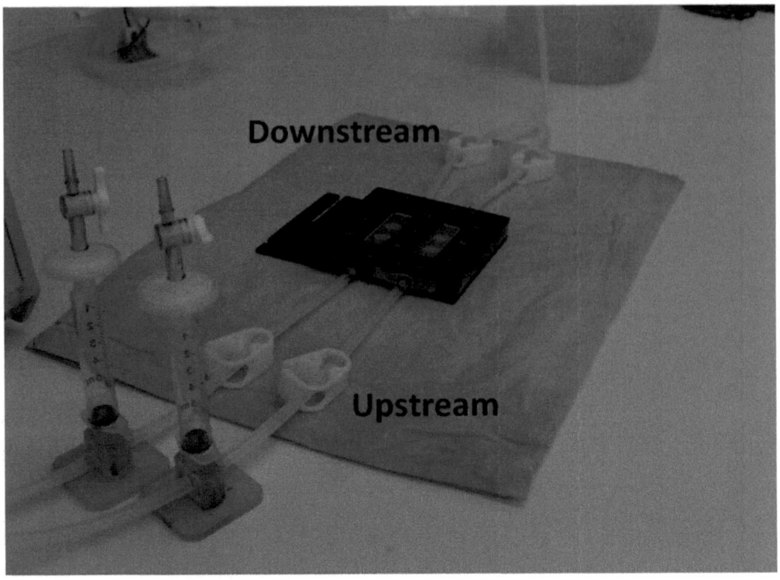

Fig. 4 Complete assembly of tubing attached to FC270 flow-cell apparatus

4. Put 17 screws into the holes and tighten until close to snug, but not completely tight. Screws must be loose to allow for material expansion during autoclave process. Complete assembly as shown in Fig. 3b.

3.2.2 Tubing Assembly of FC270

1. To assemble the tubing upstream of FC270, connect 18″ of 1/8″ tubing to a "Y" manifold, then attach two pieces of 3.18 mm pump tubing to other ends of the "Y" manifold. Next, connect straight connectors in between the ends of the 3.18 mm pump tubing and 7–8″ of 1/8″ tubing. To the ends of the 1/8″ tubing, connect the FC33 bubble traps, followed by another 5–6″ of 1/8″ tubing. Lastly, slide ratchet clamps over each 5–6″ piece and attach tubing to input ports of FC270 (Fig. 4).
2. To assemble tubing downstream of FC270, cut 18″ of 1/8″ tubing and attach to a "Y" manifold. Cut two 5–6″ pieces of 1/8″ tubing and attach to the other ends of "Y" manifold. Slide one ratchet clamp over each 5–6″ piece of tubing and attach each to the two effluent ports of FC270 (Fig. 4).
3. Cover the unattached ends of tubing and bubble traps with autoclavable paper or aluminum foil, and autoclave for 20 min on a slow vent cycle.

3.2.3 Media and Carboy Preparation

1. Prepare 4 l of biofilm media (as described in materials) in a single carboy.
2. Attach a syringe filter to the 2″ piece of 1/8″ tubing and connect to an access port on the vent-side of carboy cap.

Fig. 5 A small (2″) piece of 1/8″ tubing is attached to a 0.2 μm syringe filter connected to one of the ventilation ports on the cap of a Nalgene carboy. Inside the cap, 18″ of 1/8″ tubing is attached to the opposite ventilation port. Cover the uncovered ventilation port with aluminum foil or autoclavable paper and then autoclave

Cover the other access port with aluminum foil on the vent-side and attach 18″ of 1/8″ tubing to the inside of covered port (Fig. 5).

3. Feed tubing into carboy containing biofilm media and screw cap on to carboy. Autoclave for at least 20 min on a liquid slow vent cycle.

3.2.4 Priming the FC270 Flow-Cell System

1. Transfer the autoclaved assembly to the incubator or temperature-controlled room set at 37 °C. Tightly screw down all screws on the FC270. Lie out and adjust tubing assembly so it will not kink or bind during flow.
2. Remove aluminum foil from tubing ends and aseptically attach 1/8″ tubing to the ventilation ports on the caps of the corresponding carboys (one upstream with the biofilm media (Fig. 5) and another downstream as a waste container).
3. Screw on 0.2 μm syringe filters to the ends of two 5 ml syringes. Remove the plungers from 5 ml syringes and push syringes onto bubble traps. Attach syringe valves to the end of 5 ml syringe filters, and leave the valves open (Fig. 6).
4. Attach the flow cell tubing to the peristaltic pump by stretching 3.18 mm pump tubing tightly around the pump head and sliding

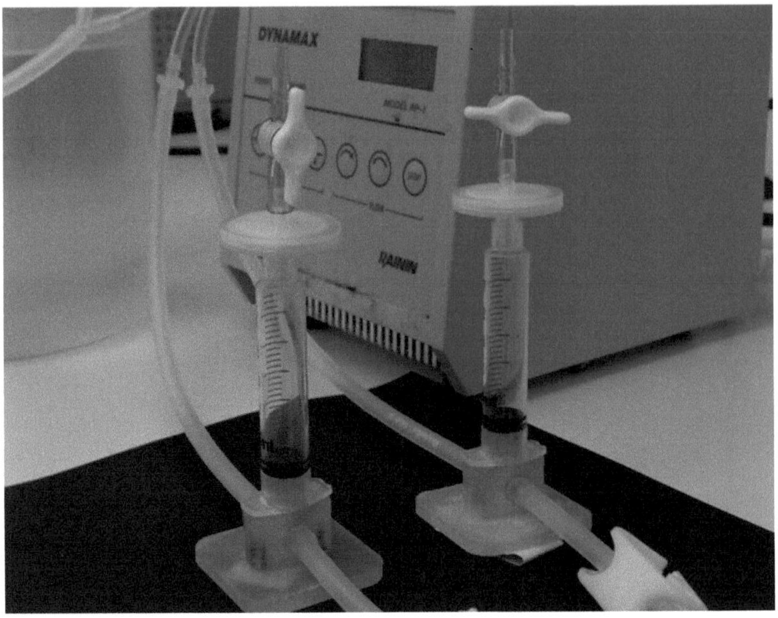

Fig. 6 Attach two 5 ml syringes to two 0.2 μm syringe filters and connect valves to the top of syringe filters. Remove plungers from 5 ml syringes and place on top of bubbles traps. Left valve is shown in "open" position and right valve shown in "closed" position

tubing securely into the front catches of the pump, making sure black and white plastic holders connected to the tubing are snug against the front catches of the pump. Tubing should fit into the indentations of the pump latches and then be latched down tightly (Fig. 7).

5. Pinch ratchet clamp tubing between the bubble traps and FC270, and begin flow of media (0.25 ml per min). Allow media to fill into syringes on the bubble traps until they are about 3–4 ml full. After syringes are 3–4 ml full, stop flow and close valves. Release ratchet clamp between bubble traps and FC270. Pump media for a couple of minutes, checking for bubbles in the system. If bubbles are present, try to remove them by gently tipping or shaking the tubing or FC270.

3.2.5 Inoculation and Starting Flow

1. To prepare inoculums, dilute overnight cultures 1:2,000 in at least 3 ml of TSB (*see* **Note 6**).
2. Draw up 3 ml of the diluted overnight culture into a 3 ml syringe with a needle. Stop flow and pinch ratchet clamp between the bubble trap and FC270. Sterilize injection site by wiping down the tubing directly upstream of FC270 with ethanol and carefully insert the needle of a 3-ml syringe into the lumen of tubing and inject adjusted inoculums (Fig. 8). Sterilize the injection site with ethanol and cover the site with

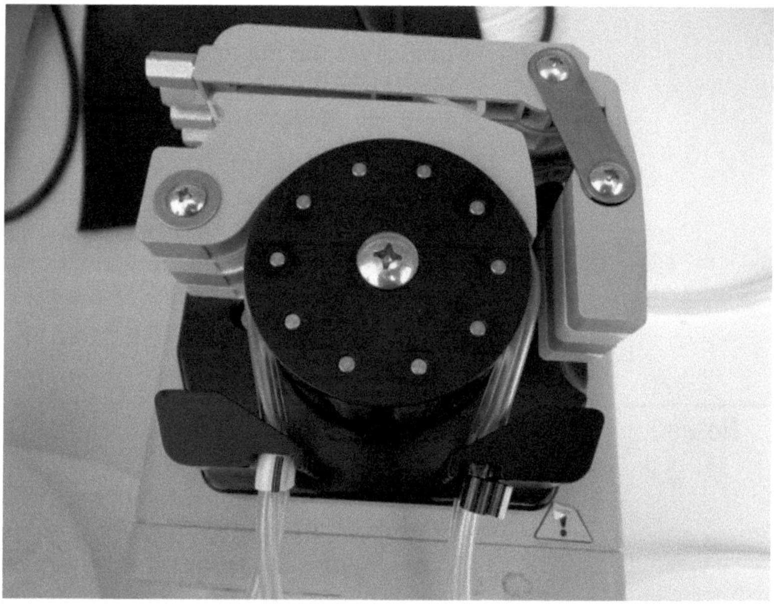

Fig. 7 The flow cell is attached to the peristaltic pump by stretching 3.18 mm pump tubing tightly around pump head and sliding tubing securely into front catches of pump, making sure black and white plastic holders connected to the tubing are snug against front catches of pump. Tubing should fit into indentations of clamps and then clamped down tightly

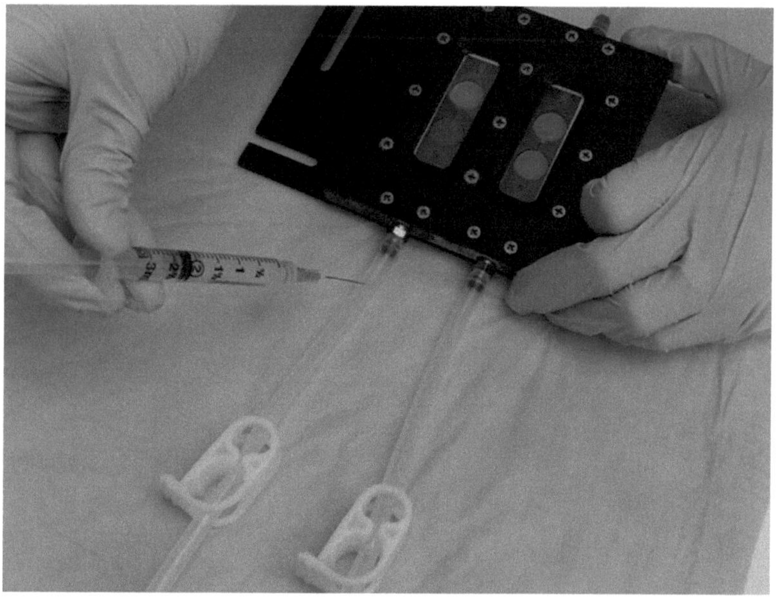

Fig. 8 The needle of 3 ml syringe full of the adjusted inoculum is carefully inserted into the lumen of the pump tubing directly upstream of the FC270 that has been wiped down with ethanol. The full inoculum is carefully injected into the flow-cell chamber, while the ratchet clamp directly upstream of inoculation site is pinched shut and the downstream ratchet clamp remains open. After full inoculum has been injected, the downstream clamp is pinched shut and injection site should be cleaned with ethanol and silicon glue applied to keep tube from leaking

silicone glue so tubing does not leak. Clamp tubing downstream of FC270 and incubate at 37 °C for 1 h to allow cells to adhere. Check for leaks in the tubing and FC270 apparatus during this time (*see* **Note 7**).

3. After 1 h, unclamp the ratchet clamps upstream and downstream of FC270. If no bubbles or leaks are found, begin pumping media at 0.25 ml/min for 24 h (*see* **Note 8**). After 24 h, stop flow and clamp on both sides of FC270, cut off excess tubing on outsides of clamps (*see* **Note 9**). The biofilm is now ready to visualized.

4 Notes

1. 0.8 OD_{600} is standard for our experimental strains. However, strain variability may require some standardization to determine an optical density that will result in more robust biofilm growth.
2. When adding liquid to the BioFlux plate, it is important to avoid bubbles in the liquid. Large bubbles can be pumped to the inside of the channels and can cause inconsistent flow, clogging, and disassociation of biofilms during an experiment.
3. BioFlux plates are constructed of glass on the bottom and are extremely fragile. It is important when handling the plates to never apply too much pressure to the bottom, especially when attaching the BioFlux interface, as they can crack allowing media to leak out of the channels and wells.
4. For complete details regarding how to operate the BioFlux Montage Software, refer to the BioFlux1000 Operation Manual. Several different experimental parameters can be adjusted (including shear rate, time, acquisition intervals, exposure of fluorescence, etc.) and should be determined and adjusted according to strain used and the objectives of the experiment.
5. When placing the cover on the top base, do not shift the cover and gasket laterally or coverslips may shift out of the indentions and break during autoclaving or when tightening the screws.
6. For *S. epidermidis* strain 1457, we have determined this inoculum to be sufficient for robust biofilm growth. However, variability in the strains used requires experimental standardization to determine the working concentrations of cells needed for optimal biofilm growth.
7. Bubbles and/or leaks in the tubing or the FC270 system can inhibit biofilm growth. Leaks can be fixed with a sealant that

will not corrode the parts of the assembly, and bubbles can often be diminished by shaking or tipping parts of the assembly. If leaks and bubbles become too problematic, disassemble the flow-cell system and fix or replace the problematic parts.

8. The flow rate was determined by collecting the effluent and measuring the amount of media over a period of time.
9. If stains or other fluorescent probes are to be introduced into the biofilm before imaging, then they must be injected at this time (similar to the inoculum) and allowed sufficient time of incubation before being imaged.

References

1. Costerton JW, Stewart PS, Greenberg EP (1999) Bacterial biofilms: a common cause of persistent infections. Science 284: 1318–1322
2. Clutterbuck AL, Woods EJ, Knottenbelt DC et al (2007) Biofilms and their relevance to veterinary medicine. Vet Microbiol 121:1–17
3. Lewis K (2001) Riddle of biofilm resistance. Antimicrob Agents Chemother 45: 999–1007
4. Hornef MW, Wick MJ, Rhen M et al (2002) Bacterial strategies for overcoming host innate and adaptive immune responses. Nat Immunol 3:1033–1040
5. Rogers KL, Fey PD, Rupp ME (2009) Coagulase-negative staphylococcal infections. Inf Dis Clin North Am 23:73–98
6. Mann EE, Rice KC, Boles BR et al (2009) Modulation of eDNA release and degradation affects *Staphylococcus aureus* biofilm maturation. PLoS One 4:e5822
7. Heydorn A, Nielsen AT, Hentzer M et al (2000) Quantification of biofilm structures by the novel computer program COMSTAT. Microbiology 146(Pt 10):2395–2407
8. Benoit MR, Conant CG, Ionescu-Zanetti C et al (2010) New device for high-throughput viability screening of flow biofilms. Appl Environ Microbiol 76:4136–4142
9. Moormeier DE, Endres JL, Mann EE et al (2013) Use of microfluidic technology to analyze gene expression during *Staphylococcus aureus* biofilm formation reveals distinct physiological niches. Appl Environ Microbiol 79:3413–3424

Chapter 14

Rapid Quantitative and Qualitative Analysis of Biofilm Production by *Staphylococcus epidermidis* Under Static Growth Conditions

Elaine M. Waters, Hannah McCarthy, Siobhan Hogan, Marta Zapotoczna, Eoghan O'Neill, and James P. O'Gara

Abstract

Rapid screening of biofilm forming capacity by *Staphylococcus epidermidis* is possible using in vitro assays with 96-well plates. This method first developed by Christensen et al. in 1985 is fast and does not require specialized instruments. Thus, laboratories with standard microbiology infrastructure and a 96-well plate reader can easily use this technique to generate data on the biofilm phenotypes of multiple *S. epidermidis* strains and clinical isolates. Furthermore, this method can be adapted to gain insights into biofilm regulation and the characteristics of biofilms produced by different *S. epidermidis* isolates. Although this assay is extremely useful for showing whether individual strains are biofilm-positive or biofilm-negative and distinguishing between form weak, moderate or strong biofilm, it is important to acknowledge that the absolute levels of biofilm produced by an individual strain can vary significantly between experiments meaning that strict adherence to the protocol used is of paramount importance. Furthermore, measuring biofilm under static conditions does not generally reflect in vivo conditions in which bacteria are often subjected to shear stresses under flow conditions. Hence, the biofilm characteristics of some strains are dramatically different under flow and static conditions. Nevertheless, rapid measurement of biofilm production under static conditions is a useful tool in the analysis of the *S. epidermidis* biofilm phenotype.

Key words Static biofilm assay, 96-well plate, Biofilm dispersal, Conditioning film, Extracellular matrix proteins

1 Introduction

Several systems have been developed to study biofilm mode of growth in *Staphylococcus sp.* Making a decision on how to grow and monitor growth is driven by the research questions that are being asked, the technology available and the costs. Fast and reproducible quantification of biofilm production by *Staphylococcus epidermidis* strains has underpinned the rapid expansion of knowledge of this important virulence determinant. One of the first methods used to achieve this was described by Christensen et al. [1]. This technique

Paul D. Fey (ed.), *Staphylococcus Epidermidis: Methods and Protocols*, Methods in Molecular Biology, vol. 1106,
DOI 10.1007/978-1-62703-736-5_14,

provides a semiquantitative measurement of biofilm production by staphylococcal cultures grown under static conditions in 96-well plates. In general this is an inexpensive and rapid method for quantifying biofilm production by multiple strains under different growth conditions, which is most useful in experiments comparing strains that form no biofilm, weak biofilm, moderate biofilm or strong biofilm. However, it is limited by significant variation between experiments in absolute levels of biofilm produced by individual strains and the fact that biofilm formation under static growth conditions does not always reflect the in vivo milieu in which cells are exposed to shear stress under flow conditions. Nevertheless, it remains an important technique for studying biofilm production and can be adapted to study the role of environmental conditions in controlling biofilm formation [2, 3], the composition of biofilm matrices [4] and the influence of the surface on which the biofilm grows to the biofilm phenotype [5].

2 Materials

2.1 Microtiter Plate Biofilm Assay

Brain heart infusion (BHI) or Tryptic Soy Broth (TSB) and BHI and TSB agar plates.

Sterile distilled water.

Spectrophotometer.

96-well plate reader with 492 nm filter.

Phosphate buffered saline.

Tissue-culture-treated polystyrene 96-well plates with lids (Nunc).

Untreated polystyrene 96-well plates with lids (Sarstedt).

Incubator (37 °C).

Drying oven (65 °C).

8-channel pipette.

NaCl.

Ethanol.

Glucose.

2.2 Congo Red Agar

BHI (Oxoid).

Microbiological agar.

Sucrose (50 % stock solution).

Congo red (0.8 mg/ml stock solution).

2.3 Biofilm Dispersal

Sodium metaperiodate.

Trypsin.

Proteinase K.

DNase I.

Dispersin B (Kane Biotech).

2.4 Biofilm Formation on Surfaces Conditioned with Matrix Proteins

S-Monovette blood collection system (Sarstedt).

15 ml test tubes.

Centrifuge.

Blood plasma prepared from fresh whole blood.

Fibronectin (Calbiochem) or other extracellular matrix proteins as required.

Bovine serum albumin.

Untreated polystyrene 96-well plates with lids (Sarstedt).

Phosphate buffered saline.

2.5 Primary Attachment Assays

Brain Heart Infusion (BHI) or Tryptic Soy Broth (TSB) and BHI and TSB agar plates.

Sterile distilled water.

Phosphate buffered saline.

Incubator (37 °C).

Tissue-culture-treated petri dishes (Nunc).

2.6 Primary Attachment and Biofilm Formation on Coupons

Brain Heart Infusion (BHI) or Tryptic Soy Broth (TSB) and BHI and TSB agar plates.

Sterile distilled water.

Phosphate buffered saline.

Sonicating water bath.

Incubator (37 °C).

3 Methods

3.1 The Christensen 96-Well Plate Biofilm Assay

The biofilms are grown in 96-well plates, which can be manufactured from either tissue-culture-treated polystyrene, which is relatively hydrophilic or untreated polystyrene, which is more hydrophobic [5]. In order to ensure reproducibility, the following procedure should be adopted (*see* **Note 1**).

1. A freezer stock of the strain to be tested should be used to inoculate BHI or TSB agar plates, which are grown at 37 °C for 24 h.
2. Three independent, single colonies should then be inoculated into 5–10 ml of BHI or TSB broth and again grown at 37 °C for 24 h.

3. These 24 h cultures should then be diluted 1:200 in fresh TSB or BHI media (supplemented as required with 4 % NaCl, 4 % ethanol, 1 % glucose, sub-inhibitory antibiotic concentrations, etc. as appropriate).
4. 100 μl of this suspension is used to inoculate the microtiter plate wells prior to aerobic or anaerobic incubation at the selected temperature(s). Biofilm production can be terminated at different time points allowing comparison biofilm development over time.
5. The microtiter plates are then washed to remove unattached and planktonic cells. Different research groups employ different methods to wash the plates. In our laboratory, the plates are washed vigorously three times by submerging the entire plate in distilled water and then shaking out the liquid. Other research groups use multichannel pipettes to decant the media from each well and replace with water or buffer in a procedure that is also repeated three times (*see* **Note 2**).
6. After washing, the plates are dried for 1 h at 65 °C prior to staining for 10 min with a 0.4 % crystal violet solution. After staining the plates are again washed three times with sterile H_2O. The absorbance of the adhered, stained cells is measured at A_{492} using a plate reader. Alternatively the crystal violet absorbed by the biofilm can be solubilized in 100 μl of 5 % acetic acid for 1 h, prior to transfer to a fresh 96-well plate and measurement of A_{492}. This latter modification can be useful in situation where biofilm formation in the well is above the dynamic range of the instrument (the extracted crystal violet can be diluted before measurement), or where the biofilm is patchy giving rise to substantial well-to-well variation, which sometimes over- or underestimates the biofilm forming capacity of the strain being measured.
7. Within each experiment, 8–16 wells are inoculated per strain and the average absorbance and standard deviation for each strain calculated. Strains are typically considered to be biofilm-positive when the average A_{492} is greater than 0.17.
8. This experiment should be repeated at least three times to generate reliable data.

3.2 Congo Red Agar Media as an Indicator of PIA-Dependent Biofilm Production by *S. epidermidis*

A helpful adjunct to using the Christensen biofilm assay can be the inspection of colony morphology of *S. epidermidis* strains grown on Congo red agar (CRA). This serves as a rapid indicator of polysaccharide intercellular adhesion (PIA) production. PIA-positive strains produce dark, dry, crusty colonies with irregular morphology on CRA, whereas PIA-negative strains produce smooth, circular, red colonies [6]. Colony morphology on CRA correlates well with PIA-dependent biofilm production in the Christensen assay.

CRA is composed of BHI agar supplemented with 0.8 mg/ml Congo red and 5 % sucrose. To prepare 100 ml of Congo red agar:

1. Add 80 ml of water to 3.7 g of BHI powder and 1 g of agar. Sterilize by autoclaving and cool to 55 °C.
2. Then add 10 ml of 50 % sucrose (pre-warmed to 55 °C) and 10 ml of 8 mg/ml Congo red (*see* **Note 3**). Mix carefully and pour plates.
3. CRA plates inoculated with *S. epidermidis* strains are incubated for 24 h at 37 °C and up to a further 72 h at room temperature.

3.3 Evaluation of the S. epidermidis Biofilm Extracellular Matrix Using Sodium Metaperiodate, Trypsin, Proteinase K, DNase I, and Dispersin B

S. epidermidis biofilm adhesins include polysaccharide intercellular adhesin (PIA), surface and extracellular proteins and extracellular DNA (eDNA). The Christensen biofilm assay can be adapted to investigate the relative contribution of these adhesins to the biofilm phenotype of specific *S. epidermidis* strains. Thus, treatment of biofilms with proteinase K or trypsin will disperse protein adhesin mediated biofilms [7, 8], dispersin B is an enzyme that degrades polysaccharide intercellular adhesin-dependent biofilms [9], sodium metaperiodate is a strong oxidant that also disperses polysaccharide biofilms [7], and DNase I [10] will disperse biofilms in which extracellular DNA is a significant matrix component. These experiments can be performed in two ways:

(a) Biofilms can be grown in 96-well plates as described above before being treated with proteinase K, trypsin, DNase I, dispersin B, or sodium metaperiodate. This approach helps determine the contribution of different biofilm adhesins to the architecture of mature biofilms.

1. Standard Christensen biofilm assays are set-up and the biofilm plates incubated for 22 h.
2. Using concentrated stock solutions, sodium metaperiodate (final concentration 10 mM), proteinase K (final concentration 10 mg/ml), trypsin (final concentration 100 μg/ml), dispersin B (final concentration 10 μg/ml), or DNase I (final concentration 0.2–1 mg/ml) are then added directly to the individual biofilm wells and mixed gently with a pipette.
3. The plates are then incubated for a further 2 h before being washed and stained as described above.

   To ensure the reliability of the data generated from these experiments it is important to include untreated controls and to include strains with known biofilm phenotypes e.g. *S. epidermidis* 1457 or RP62A, which are known to produce PIA-type biofilms under standard growth conditions. Thus, mature 1457 or RP62A biofilms grown in BHI or TSB media will be significantly dispersed by dispersin B or sodium metaperiodate. In contract proteinase K or DNase I will have no significant effect on these PIA biofilms but will disperse biofilms in which surface or extracellular proteins are the most important adhesins.

(b) In the second approach proteinase K, trypsin, DNase I, or dispersin B can be added to the cultures prior to incubation (sodium metaperiodate inhibits growth and cannot be used in this way). Biofilms are again grown for 24 h before being washed as described above.

This method allows the contribution of different biofilm adhesins to the development of a growing biofilm to be investigated. Thus, biofilm production by *S. epidermidis* strains dependent on surface or extracellular proteins, eDNA, or PIA will be inhibited by proteinase K, DNase I, or dispersin B, respectively.

3.4 Biofilm Formation on Polystyrene Surfaces Conditioned with Host Extracellular Matrix Proteins

Implanted biomaterial susceptible to *S. epidermidis* biofilm infections are rapidly coated by a conditioning file of host extracellular matrix proteins. The interaction of *S. aureus* with these proteins often plays an important role in the initiation of biofilm production [11–13]. In contrast, biofilm formation by *S. epidermidis* and other species of coagulase negative staphylococci is generally not dependent on a conditioning film, marking an important biofilm difference between these species. Immobilization of blood plasma proteins to polystyrene is an effective way to establish a clinically relevant conditioning film. The approach used by our group is as follows:

1. Blood (approximately 40 ml) is drawn from a volunteer using the S-Monovette system (Sarstedt), which is a kit containing pre-made syringes with heparin coated beads.
2. The blood is gently transferred into 15 ml tubes for separation into platelet rich plasma (PRP) by centrifugation at 150 × g for 10 min. Pure plasma can be obtained by centrifugation of whole blood or PRP at 1,000–2,000 × g for 10 min.
3. After centrifugation, plasma is the upper layer which should be collected gently and used immediately or frozen at −20 °C. Roughly 40 % of blood drawn can be obtained back as plasma.
4. A range of plasma concentrations (from 25 %—neat) can be used to condition the polystyrene plates. The plasma is diluted in carbonate buffer pH 9.6, which promotes immobilization of the matrix protein to the polystyrene plates.
5. A volume of 100 μl of the plasma solution is added to the individual wells of a tissue-culture-treated 96-well plate and incubated at 37 °C for 2 h.
6. After immobilization of the plasma proteins to the 96-well plates, biofilm assays can be set up as described in Subheading 3.1.

This assay can also be adapted to investigate the role of individual matrix proteins such as fibronectin, as follows:

1. Doubling dilutions of human fibronectin (0–20 μg/ml) are added in 100 μl volumes to untreated polystyrene 96-well plates and incubated overnight at 4 °C.
2. The microtiter plates are then washed three times in sterile PBS and blocked for 2 h at 37 °C with 5 % bovine serum albumin.

Other matrix proteins can be immobilized to polystyrene using this method. Biofilm assays are then set up as described previously.

3.5 Measurement of Primary Attachment by *S. epidermidis* Cells to Surfaces

Biofilm formation by *S. epidermidis* is typically considered to be a two step process initiated by the attachment of a planktonic cell to a surface followed by the accumulation of the bacterial cells into a biofilm. A number of methods have been developed to quantify the adhesiveness or primary attachment of *S. epidermidis* cells to surfaces. These techniques are based on inoculating a defined number of cells onto a surface, incubating for 30–60 min before washing off unattached cells and quantifying the number of attached cells.

The primary attachment assay used in our laboratory [4] is based on the method of Lim et al. [14].

1. A freezer stock of the strain to be tested should be used to inoculate BHI or TSB agar plates, which are grown at 37 °C for 24 h.
2. Three single colonies should then be inoculated into 5–10 ml of BHI or TSB broth and again grown at 37 °C for 24 h.
3. The overnight cultures are diluted in the same media to give approximately 300 CFU/100 μl, as determined by serial dilution and spread plating on TSA or BHI agar plates.
4. 100 μl of the adjusted cell suspension is then spread onto a Nunclon tissue-culture-treated (Δ surface) petri dishes. As noted previously tissue-culture-treated polystyrene is more hydrophilic than untreated polystyrene and promotes cell attachment.
5. After incubation at 37 °C for 30 min, the Petri dishes are rinsed gently three times with 5 ml of sterile PBS (pH 7.5) and then covered with 15 ml of molten 0.8 % BHI cooled to 48 °C. The molten agar is allowed to solidify and the Petri dishes are then incubated overnight at 37 °C.
6. Cells that remain attached to the polystyrene after washing give rise to colonies on the underside of the agar and can be counted. Primary attachment is expressed as the percentage of CFU remaining on the petri dishes after washing.
7. Each experiment is repeated at least three times.

3.6 Measurement of Primary Attachment and Biofilm Accumulation on Coupons

There remains considerable interest in the development of novel biomaterials and surface coatings, which are less susceptible to colonization and biofilm-associated infections. The potential of *S. epidermidis* to form biofilm on such surfaces cannot be assessed using the microtiter plate assay. However, small coupons with a defined surface area (e.g. 1 cm^2) cut from these materials can be used as a template on which attachment and biofilm formation can be measured. If required, the coupons can be immersed in blood plasma or extracellular matrix proteins prior to incubation with the bacteria as follows.

1. 1 cm^2 coupons are placed in 24-well plates and 1 ml of blood plasma or ECM protein solution added to the well before incubation at 37 °C for 2 h or 4 °C for 24 h, respectively.
2. The coupons are then gently rinsed three times in phosphate buffered saline and blocked for 2 h at 37 °C with 5 % bovine serum albumin before being again rinsed three times in PBS and transferred to fresh 24-well plates.
3. Overnight broth cultures (prepared as described for the Christensen 96-well plate assay) are adjusted to $A_{600} = 1.0$. The number of colony-forming units in the inocula is determined by preparing serial dilutions and inoculating onto BHI or TSB agar.
4. 1 ml of the $A_{600} = 1$ inocula is added to each of the coupons in the individual well of a 24-well plate and incubated at 37 °C for 1 h to allow bacterial cell attachment.
5. The coupons are then removed and gently washed in a new 24-well plate containing 3 ml of sterile dH_2O on a shaker for 2 min at 200 rpm to remove any unattached planktonic cells.
6. Each coupon is then placed in a sterile universal bottle contain 1 ml of sterile dH_2O, vortexed for 5 min, sonicated for 2 min in ultrasonic water bath, and finally vortexed again for another 5 min. Gentle sonication assists in detaching the biofilm from the surface of the coupon, and vortexing assists in disruption of the biofilm aggregates to generate a homogeneous cell suspension. However, undoubtedly cell aggregates remain meaning that this method is best suited to comparing relative cell attachment by the same strain to different types of coupons.
7. The cells removed from the coupon surface are serially diluted and spread plated on BHI agar before being incubated at 37 °C for 24 h. The number of CFU recovered from the coupon surface is expressed as a percentage (% attachment) of the number of CFU in the inoculum.

To measure biofilm formation

1. 3 ml of the $A_{600} = 1$ inoculum is added to the coupons in the individual wells of a 24-well plate and incubated at 37 °C for 24 h.

2. The coupons are then removed, rinsed in distilled H_2O and transferred to a universal bottle containing 9 ml of sterile PBS. The coupons are then vortexed for 5 min, sonicated in a water bath for 2 min, and then vortexed again for 2 min.
3. The cell suspension is then serially diluted and plated on BHI agar to determine the number of colony-forming units. The data are then expressed as the log of the number of colony-forming units per surface area of coupon (e.g. log cfu/cm^2).

4 Notes

1. Although the general biofilm characteristics of a strain can be reliably determined by the Christensen method, there is often significant experiment to experiment variation in terms of the exact levels of biofilm produced by individual strains. This variation can be attributed in part to different batches of media, incubation temperatures, and different methods of inocula production. Therefore, where possible, it is best to start each experiment using the same freezer stock and the same batch of media. Furthermore, it is important to repeat the experiment at least three times for each strain.
2. The microtiter plate biofilm assay can also be adapted to measure the viability of biofilms treated with antimicrobial drugs using Alamar Blue (resazurin), a nonfluorescent redox dye, which is reduced in the presence of active microbial metabolism to fluorescent resorufin. After growing biofilms in microtiter plates for 24 h, the plates are washed twice with sterile water and a 25 % solution of the Alamar Blue (Biosource, Invitrogen) added to the wells and incubated in the dark for 60 min at 37 °C. The production of a pink color is indicative of redox activity and viability, whereas a blue color indicates a lack of viability. The degree of biofilm viability can be quantified using a fluorimeter set at 560 nm excitation and 585 nm emission wavelengths. Each experiment is carried out a minimum of three times.
3. The Congo red dye has a tendency to precipitate out of solution at room temperature and should be heated to 60 °C prior to each use.

References

1. Christensen GD, Simpson WA, Younger JJ et al (1985) Adherence of coagulase-negative staphylococci to plastic tissue culture plates: a quantitative model for the adherence of staphylococci to medical devices. J Clin Microbiol 22:996–1006
2. Conlon KM, Humphreys H, O'Gara JP (2002) *icaR* encodes a transcriptional repressor involved in environmental regulation of *ica* operon expression and biofilm formation in *Staphylococcus epidermidis*. J Bacteriol 184: 4400–4408

3. Conlon KM, Humphreys H, O'Gara JP (2004) Inactivations of *rsbU* and *sarA* by IS256 represent novel mechanisms of biofilm phenotypic variation in *Staphylococcus epidermidis*. J Bacteriol 186:6208–6219
4. Holland LM, Conlon B, O'Gara JP (2011) Mutation of *tagO* reveals an essential role for wall teichoic acids in *Staphylococcus epidermidis* biofilm development. Microbiology 157: 408–418
5. Kennedy CA, O'Gara JP (2004) Contribution of culture media and chemical properties of polystyrene tissue culture plates to biofilm development. J Med Microbiol 53: 1171–1173
6. Handke LD, Conlon KM, Slater SR et al (2004) Genetic and phenotypic analysis of biofilm phenotypic variation in multiple *Staphylococcus epidermidis* isolates. J Med Microbiol 53:367–374
7. Mack D, Siemssen N, Laufs R (1992) Parallel induction by glucose of adherence and a polysaccharide antigen specific for plastic-adherent *Staphylococcus epidermidis*: evidence for functional relation to intercellular adhesion. Infect Immun 60:2048–2057
8. Rohde H, Burandt EC, Siemssen N et al (2007) Polysaccharide intercellular adhesin or protein factors in biofilm accumulation of *Staphylococcus epidermidis* and *Staphylococcus aureus* isolated from prosthetic hip and knee joint infections. Biomaterials 28:1711–1720
9. Kaplan JB, Ragunath C, Velliyagounder K et al (2004) Enzymatic detachment of *Staphylococcus epidermidis* biofilms. Antimicrob Agents Chemother 48:2633–2636
10. Kaplan JB, Izano EA, Gopal P et al (2012) Low levels of beta-lactam antibiotics induce extracellular DNA release and biofilm formation in *Staphylococcus aureus*. mBio 3:e00198–00112
11. O'Neill E, Pozzi C, Houston P et al (2008) A novel *Staphylococcus aureus* biofilm phenotype mediated by the fibronectin-binding proteins, FnBPA and FnBPB. J Bacteriol 190: 3835–3850
12. Geoghegan JA, Monk IR, O'Gara JP et al (2013) Subdomains N2N3 of fibronectin binding protein A mediate *Staphylococcus aureus* biofilm formation and adherence to fibrinogen using distinct mechanisms. J Bacteriol 195:2675–2683
13. Beenken KE, Blevins JS, Smeltzer MS (2003) Mutation of *sarA* in *Staphylococcus aureus* limits biofilm formation. Infect Immun 71: 4206–4211
14. Lim Y, Jana M, Luong TT et al (2004) Control of glucose- and NaCl-induced biofilm formation by *rbf* in *Staphylococcus aureus*. J Bacteriol 186:722–729

Chapter 15

Bacteriophage Transduction in *Staphylococcus epidermidis*

Michael E. Olson and Alexander R. Horswill

Abstract

The genetic manipulation of *Staphylococcus epidermidis* for molecular experimentation has long been an area of difficulty. Many of the traditional laboratory techniques for strain construction are laborious and hampered by poor efficiency. The ability to move chromosomal genetic markers and plasmids using bacteriophage transduction has greatly increased the speed and ease of *S. epidermidis* studies. These molecular genetic advances have advanced the *S. epidermidis* research field beyond a select few genetically tractable strains and facilitated investigations of clinically relevant isolates.

Key words *Staphylococcus epidermidis*, Bacteriophage, Transduction, Bacteriophage 71

1 Introduction

Transduction is the process by which DNA is transferred from one bacterium to another by bacterial viruses known as bacteriophage. Norton Zinder and Joshua Lederberg discovered this process in 1951 [1]. Bacteriophage reproduction relies on replicational, transcriptional, and translation machinery of the host bacterial cell to make new virions. During this process, the packaging of bacteriophage DNA is a relatively low-fidelity event, and pieces of bacterial chromosome may become mistakenly packaged into the bacteriophage capsid. The lytic cycle leads to a bloom of new bacteriophage particles, which are released by lysis of the host cell [2, 3]. Research methods for bacterial strain construction have taken advantage of the accidental bacterial DNA packaging into the bacteriophage to manipulate recipient strains by selection of antibiotic resistance markers or other genetic determinants.

Generalized transduction is the process by which any bacterial gene may be transferred to another bacterium via a bacteriophage. This is in contrast to specialized transduction, which is restricted on the DNA the bacteriophage can move. Generalized transduction

Paul D. Fey (ed.), *Staphylococcus Epidermidis: Methods and Protocols*, Methods in Molecular Biology, vol. 1106, DOI 10.1007/978-1-62703-736-5_15, © Springer Science+Business Media, LLC 2014

occurs by a headful packaging mechanism, followed by infection of the recipient bacterial host and recombination onto the chromosome. The use of transduction in molecular microbiology labs relies on lytic bacteriophages, as opposed to a lysogenic cycle where the bacteriophage DNA integrates into a specific site in the host chromosome and remains dormant. During the lytic cycle of infection, the virus takes control of the bacterial cell's machinery to replicate its own DNA and produce more viral particles. In headful packaging, the bacteriophages fill their capsid with viral genetic material, but the low-frequency accidental packing of bacterial plasmid or chromosomal DNA into the viral capsid sets the stage for downstream generalized transduction. Following another round of lytic infection, the bacteriophages infect new recipient bacteria and inject the foreign DNA (viral and bacterial) into the recipient cells. In this transduction event, the transferred bacterial DNA can integrate into the recipient bacterium's genome through homologous recombination or recircularize into a replicating plasmid [3]. The final result is the successful transduction of bacterial genetic information from one strain to another, which is especially important as a mechanism through which antibiotic-resistance genes are exchanged between bacteria [2, 4]. Pathogenicity islands are also known to horizontally transfer to new strains using this mechanism [5].

Transduction has been adapted as a laboratory method for transferring genetic material to manipulate *S. epidermidis.* The isolation and first use of bacteriophage 71 trace back to early work on bacteriophage-typing staphylococci by Baird-Parker in the early 1960s [6, 7]. Dietrich Mack and colleagues adapted bacteriophage 71 [8] for generalized transduction in *S. epidermidis* in 1998 [9]. In order to improve transduction of chromosomal markers, bacteriophage 71 lysates were UV-irradiated [10, 11]. The addition of bacteriophage transduction to the molecular "toolbox" for *S. epidermidis* has enhanced and accelerated discoveries in clinical isolates. In this chapter, we outline a straightforward method for generalized transduction of chromosomal markers and plasmids in *S. epidermidis* using bacteriophage 71.

2 Materials

1. 13×100 mm Tryptic soy agar (TSA)/brain–heart infusion (BHI) slants.
2. Tryptic soy broth (TSB) + 5 mM $CaCl_2$.
3. Petri plates.
4. 15 mL Falcon tubes (BD Biosciences) or equivalent.
5. TSA (1.5 % agar) + 5 mM $CaCl_2$.

(a) TSA (1.5 % agar) + 500 mg/L Na citrate + antibiotic of choice.

(b) Soft agar TSA (0.5 % agar) + 5 mM $CaCl_2$.

6. 0.5 M $CaCl_2$.
7. 0.02 M Na citrate.
8. Antibiotic of choice; typical antibiotics and concentrations include erythromycin 10 μg/mL, chloramphenicol 10 μg/mL, trimethoprim 10 μg/mL, tetracycline 2–10 μg/mL, kanamycin 50 μg/mL.

2.1 Equipment

1. Centrifuge.
2. Incubators—static and shaking.
3. Water bath 50 °C.

3 Methods

3.1 Bacteriophage 71 Propagation

Transduction of plasmid and chromosomal markers of interest requires a phage titer of approximately 10^{10} pfu/mL. As phage titers gradually decrease during storage at 4 °C, it is appropriate to propagate phage 71 to acquire a high titer (10^{10}) before the transducing lysate is generated.

1. Grow *S. epidermidis* propagation strain on 13 × 100 mm TSA slant overnight at 37 °C. Ensure that a plasmid-free strain of bacteriophage 71-susceptible *S. epidermidis* is used. Strain 1457 is recommended as it is both a good recipient and propagation strain for phage 71 [12] and allows for optimal phage titers (10^{10} pfu—plaque-forming units; *see* **Note 1**).
2. Resuspend *S. epidermidis* propagation strain in 1 mL TSB + 5 mM $CaCl_2$ (*see* **Note 2**).
3. Add 4 mL TSA soft agar to ten Falcon tubes (15 mL). Hold in 50 °C water bath to prevent agar from solidifying.
4. Serially dilute bacteriophage 71 stock 10-fold to 10^{-10} in TSB + 5 mM $CaCl_2$.
5. Combine 10 μl *S. epidermidis* cells and 100 μl bacteriophage dilution to soft agar. Gently mix (do not vortex) and pour onto TSA + 5 mM $CaCl_2$ plates. Repeat for all ten bacteriophage dilutions. Fresh TSA plates work best to prevent soft agar from drying out during phage propagation.
6. Incubate overnight (plates right side up) at 37 °C. Do not invert plates to make sure that soft agar is maintained on the agar surface.

3.2 Harvest Bacteriophage and Titer Determination

1. Select up to three plates for bacteriophage harvest. Optimal plates will show near-confluent lysis and minimal bacterial growth (ideally the 10^{-3} to 10^{-5} plates, but this depends upon the original titer of bacteriophage 71 stock).
2. Add 3 mL TSB to plates. Harvest bacteriophage by breaking up and scraping off soft agar with a plate spreader. Transfer resulting agar/TSB mixture to a 50 mL tube.
3. Break up agar as much as possible by gently pipetting up and down to facilitate the release of bacteriophage particles. Avoid bubbles, vortexing, and sonication as they mechanically sheer bacteriophage tails.
4. Centrifuge for 10 min at 10,000 × *g*.
5. Filter supernatant through 0.45 μm filter.
6. Store bacteriophage at 4 °C.
7. The titer of the resulting bacteriophage lysate should be determined by repeating the experiment outlined in Subheading 3.1; optimal bacteriophage titer should be approximately 10^{10} pfu/mL. In some cases, when the original bacteriophage 71 stock titer is low, multiple propagation experiments may be required to acquire a titer of 10^{10} pfu/mL.

3.3 Preparation of Transducing Lysate

1. Repeat bacteriophage propagation and harvest protocol (3.1 and 3.2) using *S. epidermidis* strain of interest (either plasmid or chromosomal marker). Note that overnight growth may require 30 °C if using temperature-sensitive plasmid (i.e., $pE194_{ts}$ derived). 10^{10} pfu/mL of the transducing lysate should be achieved to ensure an appropriate transduction frequency (~10^{-8}).

3.4 Transduction

1. Grow the transduction recipient overnight on a 13 × 100 mm TSA slant.
2. Resuspend recipient strain in 1 mL TSB + 5 mM $CaCl_2$.
3. Add 500 μL of the recipient strain suspension to a 50 mL tube.
4. Add 1.5 mL TSB + 5 mM $CaCl_2$.
5. Add 500 μL bacteriophage 71 transducing lysate (10^{10} pfu/mL) to tube (*see* **Note 3**).
6. Shake at 225 RPM for exactly 20 min at 37 °C (30 °C if transducing a temperature-sensitive plasmid).
7. Add 1 mL cold (4 °C) 0.02 M Na citrate.
8. Centrifuge at 2,000 × *g*, 4 °C, for 10 min.
9. Resuspend pellet in 1 mL cold (4 °C) 0.02 M Na citrate.
10. Plate 100 mL of transduced cells each to TSA + 500 mg/L Na citrate + antibiotic plates.

11. Incubate at 37 °C (30 °C if transducing a temperature-sensitive plasmid).
12. Pick single colonies to streak for isolation on the TSA + 500 mg/L Na citrate + antibiotic plates (*see* **Note 4**).
13. Confirm movement of plasmid/chromosomal marker by standard methods including plasmid analysis, PCR, or Southern blot.

4 Notes

1. A rigorous examination of *S. epidermidis* strains that are susceptible to phage 71 has not been performed. If using *S. epidermidis* strains other than 1457, check susceptibility of each strain to phage 71 by first streaking each strain to be tested on TSA containing 5 mM $CaCl_2$. 10 μl of phage 71 stock (10^{10} pfu/mL) is then spotted on the plate in the first quadrant and allowed to dry. The plate is incubated at 37 °C for 24 h; an area of lysis will be evident in *S. epidermidis* phage 71-susceptible strains.
2. Calcium chloride ($CaCl_2$) is added to the media to facilitate bacteriophage 71 attachment to *S. epidermidis*. The addition of Na citrate chelates the calcium, halting the bacteriophage infectious cycle and preventing reinfection.
3. As a negative control, perform the transduction experiments using all components except the bacteriophage to control for contaminated phage transduction lysate.
4. Na citrate is required in these plates to chelate residual calcium in the TSA. Although individual transductant colonies are picked, the titer of bacteriophage 71 on the plate is so high that phages are typically transferred to subsequent TSA plates resulting in partial lysis of colony growth if Na citrate is not added.

Acknowledgements

This work was supported by project 3 of NIH grant AI083211 from the National Institute of Allergy and Infectious Diseases.

References

1. Lederberg J, Lederberg EM, Zinder ND et al (1951) Recombination analysis of bacterial heredity. Cold Spring Harbor Symp Quant Biol 16:413–443
2. Novick RP, Christie GE, Penades JR (2010) The phage-related chromosomal islands of Gram-positive bacteria. Nat Rev Microbiol 8:541–551

3. Griffiths AJF, Suzuki DT, Lewontin RC et al (2000) An introduction to genetic analysis, 7th edn. W. H. Freeman, New York
4. Christie GE, Dokland T (2012) Pirates of the caudovirales. Virology 434:210–221
5. Ruzin A, Lindsay J, Novick RP (2001) Molecular genetics of SaPI1—a mobile pathogenicity island in Staphylococcus aureus. Mol Microbiol 41:365–377
6. Baird-Parker AC (1963) A classification of micrococci and staphylococci based on physiological and biochemical tests. J Gen Microbiol 30:409–427
7. Baird-Parker AC (1965) Staphylococci and their classification. Ann N Y Acad Sci 128:4–25
8. Verhoef J, Winkler KC, van Boven CP (1971) Characters of phages from coagulase-negative staphylococci. J Med Microbiol 4:413–424
9. Nedelmann M, Sabottke A, Laufs R et al (1998) Generalized transduction for genetic linkage analysis and transfer of transposon insertions in different *Staphylococcus epidermidis* strains. Zentralbl Bakteriol 287:85–92
10. Kayser FH, Wust J, Corrodi P (1972) Transduction and elimination of resistance determinants in methicillin-resistant *Staphylococcus aureus*. Antimicrob Agents Chemother 2:217–223
11. Mack D, Nedelmann M, Krokotsch A et al (1994) Characterization of transposon mutants of biofilm-producing *Staphylococcus epidermidis* impaired in the accumulative phase of biofilm production: genetic identification of a hexosamine-containing polysaccharide intercellular adhesin. Infect Immun 62: 3244–3253
12. Mack D, Fischer W, Krokotsch A et al (1996) The intercellular adhesin involved in biofilm accumulation of *Staphylococcus epidermidis* is a linear beta-1,6-linked glucosaminoglycan: purification and structural analysis. J Bacteriol 178:175–183

Chapter 16

Mouse Model of Post-arthroplasty *Staphylococcus epidermidis* Joint Infection

Tyler D. Scherr, Kevin E. Lindgren, Carolyn R. Schaeffer, Mark L. Hanke, Curtis W. Hartman, and Tammy Kielian

Abstract

Animal models are invaluable tools for translational research, allowing investigators to recapitulate observed clinical scenarios within the laboratory that share attributes with human disease. Here, we describe a mouse model of post-arthroplasty *Staphylococcus epidermidis* joint infection which mimics human disease and may be utilized to explore the complex series of events during staphylococcal implant-associated infections by identifying key immunological, bacterial, and/or therapeutic mechanisms relevant to these persistent infections.

Key words *Staphylococcus epidermidis*, Biofilm, Post-arthroplasty, Implant-associated, Medical device, Osteomyelitis

1 Introduction

Staphylococcus epidermidis is a gram-positive bacterium that naturally colonizes the human skin. However, as an opportunistic pathogen *S. epidermidis* has become a major cause of nosocomial infections, particularly afflicting immunocompromised or immunosuppressed individuals. While *S. epidermidis* does not possess the arsenal of virulence factors associated with other leading pathogens, it has an impressive propensity to form complex, heterogeneous communities, known as biofilms, primarily on artificial surfaces [1]. As such, *S. epidermidis* is widely recognized as the leading cause of implanted medical device infections [2, 3]. To date, *S. epidermidis* has been associated with a wide range of device infections, including intravenous catheters, cerebral spinal fluid shunts, heart valves, pacemakers, prosthetic joints, and intramedullary nails [4–7]. These biofilm-associated infections are typically recalcitrant to antibiotic therapy as a result of altered metabolism during biofilm growth [8–11]. Ultimately, surgical removal and

Paul D. Fey (ed.), *Staphylococcus Epidermidis: Methods and Protocols*, Methods in Molecular Biology, vol. 1106, DOI 10.1007/978-1-62703-736-5_16,

replacement of the infected device is often required, which is a long and debilitating process associated with significant morbidity and economic impact for patients [2]. With the increasing worldwide demand for medical device implants, particularly orthopedic implants associated with hip and knee arthroplasties, there is a pressing need for the development of novel therapeutics to combat *S. epidermidis* implant-associated infections [12].

While staphylococcal adherence and biofilm formation on artificial surfaces have been studied extensively in vitro, there are currently a limited number of useful animal models [13, 14]. Current animal models fall into the two basic categories of catheter-associated or orthopedic implant-associated infections. The catheter-associated models allow researchers to explore host–pathogen interactions in varying tissues and cavities throughout the body and, depending on location, provide the opportunity to examine biofilm growth under dynamic conditions. In contrast, orthopedic implant-associated models explore static biofilm growth and host–pathogen interactions in the immunologically fertile bone marrow. Furthermore, this model can be used to examine the rising incidence of implant-associated osteomyelitis, of which *S. epidermidis* is also a leading cause [14]. The following mouse model of post-arthroplasty *S. epidermidis* joint infection attempts to accurately mimic several facets of human disease seen in the clinical setting, including implant materials of chronicity of infection. Titanium alloys are the most common metals used in total-joint arthroplasties, and, as such, nickel-titanium wire was selected for the implant material in the model described here [15]. Likewise, surgical tools and sutures were selected to mimic those used clinically.

2 Materials

Prepare all solutions using sterile ultrapure water or PBS and analytical grade reagents. Prepare and store all reagents at room temperature (unless indicated otherwise). Strict adherence should be made to follow all waste disposal regulations when handling infectious hazardous waste materials.

2.1 Bacterial Preparation

1. Tryptic soy broth (TSB).
2. One liter flask(s).
3. Centrifuge and appropriate tubes (*see* **Note 1**).
4. Sterile saline or PBS (Hyclone Laboratories, Logan, UT).
5. Tubes for dilution plating.
6. TSA plates.

2.2 Bone Wire Insertion and Inoculation

1. Mouse strain: Can be any strain desired, including outbred and inbred strains. Our laboratory typically uses C57BL/6 mice, since the majority of mouse knockout strains have been generated on the C57BL/6 background. Animals typically between 8 and 10 weeks of age are used.
2. Anesthesia, 10 % ketamine (100 mg/mL) with 2.5 % xylazine (20 mg/mL) in sterile PBS. Ten milliliter ketamine/xylazine anesthesia will provide anesthetic for approximately forty 25 g mice. Sterilize the top of the ketamine vial with alcohol. Using a 1 cc syringe with an 18 gauge needle, aseptically remove 1 mL from the ketamine stock vial and dispense into a 50 mL conical tube. Using a second syringe, aseptically remove 0.25 mL xylazine from the stock tube and also dispense into the same 50 mL conical tube. Bring to a final volume of 10 mL by adding 8.75 mL of PBS, and vortex thoroughly to mix. Ketamine/xylazine can be stored at 4 °C for up to 1 week. Note that ketamine is a controlled substance and should be stored in a secure location with appropriate recordkeeping of usage.
3. Buprenex® buprenorphine HCl, 0.3 mg/mL (dilute 1:10 in sterile saline or water) (Reckitt Benckiser Pharmaceuticals Inc., Richmond, VA): Note that Buprenex® is a controlled substance and should be stored in a secure location with appropriate recordkeeping of usage.
4. Sterile surgical towels/drapes.
5. Sterile gauze.
6. Electric clippers to remove fur.
7. Puralube® vet ointment or comparable ophthalmic ointment (Fera Pharmaceuticals, Locust Valley, NY).
8. Povidone-Iodine cleansing solution, 7.5 % (CareFusion, Leawood, KS).
9. Small animal scale.
10. Mouse ear punch tool.
11. Surgical tools including Iris scissors, Adson forceps, Kelly artery forceps, scalpel and size 15 blades, and needle driver.
12. Nickel-Titanium wire (0.6 mm in diameter) (Custom Wire Technologies, Inc., Port Washington, WI). Wire should be carefully cut to a length of 8 mm and autoclaved before use.
13. 23 and 26 gauge needles.
14. 1 cc syringes.
15. Sutures: 6-0 Polysorb absorbable suture for the underlying fascia and 6-0 Monosof monofilament nylon suture to close the skin.

16. 10 μL Hamilton syringe, fitted with a 26-gauge blunt-end needle. Rinse the syringe by pulling up 10 μL of 70 % ethanol. Repeat several times. Follow by rinsing the syringe several times with sterile water or PBS to remove any trace ethanol before loading the syringe with bacteria (*see* **Note 2**).

2.3 Sacrifice Animals and Sample Collection

1. Dissection tools, including scissors, forceps.
2. Sterile PBS.
3. 5 mL polystyrene tubes.
4. Homogenization buffer: PBS supplemented with protease inhibitor cocktail tablet (#116 97 498 001; Roche Diagnostics, Indianapolis, IN). Store working aliquots at 4 °C or −20 °C for long-term storage.
5. 1.5 or 2 mL microcentrifuge tubes.
6. Small animal scale.

2.4 Sample Processing and Analysis

1. 1.5 mL microcentrifuge tubes.
2. Bullet Blender (Next Advance Inc., Averill Park, NY) and grinding beads (*see* **Note 3**).
3. Polytron tissue homogenizer (Kinematica AG, Luzern, CH).
4. TSA plates, previously prepared in **item 6** in Subheading 2.1.
5. Graphing and statistical software (*see* **Note 4**).
6. Mouse microbead array (MILLIPLEX; Millipore, Billerica, MA).
7. Bio-Plex workstation (Bio-Rad, Hercules, CA).
8. BCA protein assay (Bio-Rad, Hercules, CA).

3 Methods

3.1 Bacterial Preparation

1. Inoculate 100 mL of bacterial growth media in a 1 L flask. Incubate at 37 °C with aeration (200 rpm) for 18 h.
2. The following day, centrifuge 30 mL of the culture at 6,000 × *g* for 3 min. Bacterial cultures should be approximately 10^9 CFU/mL. The centrifugation steps concentrate the culture(s), resulting in an inoculum of approximately 10^8 CFU in 2 μL.
3. Decant the supernatant, and wash the cell pellet with 10–15 mL sterile saline or PBS. Centrifuge as above.
4. Decant the supernatant, and resuspend the cell pellet in 3 mL sterile saline or PBS. Confirm inoculum by titering the bacteria on TSA plates. Keep bacterial preparation(s) on ice until use.

3.2 Bone Wire Insertion and Inoculation

1. Weigh mouse, anesthetize by intraperitoneal injection (10 μL of ketamine/xylazine per gram body weight), and ear punch for animal identification. Leave the mouse undisturbed for 5–10 min without manipulation to ensure that it is fully anesthetized.
2. Remove all hair from the right lower extremity using the electric hair clippers. Apply ophthalmic ointment to both eyes to prevent them from drying out during the surgical procedure.
3. Place the mouse in a supine position on the operating platform and scrub the lower right extremity (leg) with povidone-iodine cleansing solution. Place a sterile isolation drape over the extremity (Fig. 1a) to provide a sterile operative field (*see* **Note 5**).
4. To begin the procedure, flex the knee and use the scalpel to make a midline incision over the knee through the skin and subcutaneous tissue, stopping at the depth of the fascia overlying the patella (Fig. 1b).
5. Perform a medial parapatellar arthrotomy with the scalpel to open the knee joint (Fig. 1c; *see* **Note 6**).
6. Laterally retract the patella to expose the entire distal femur (*see* **Note 7**).

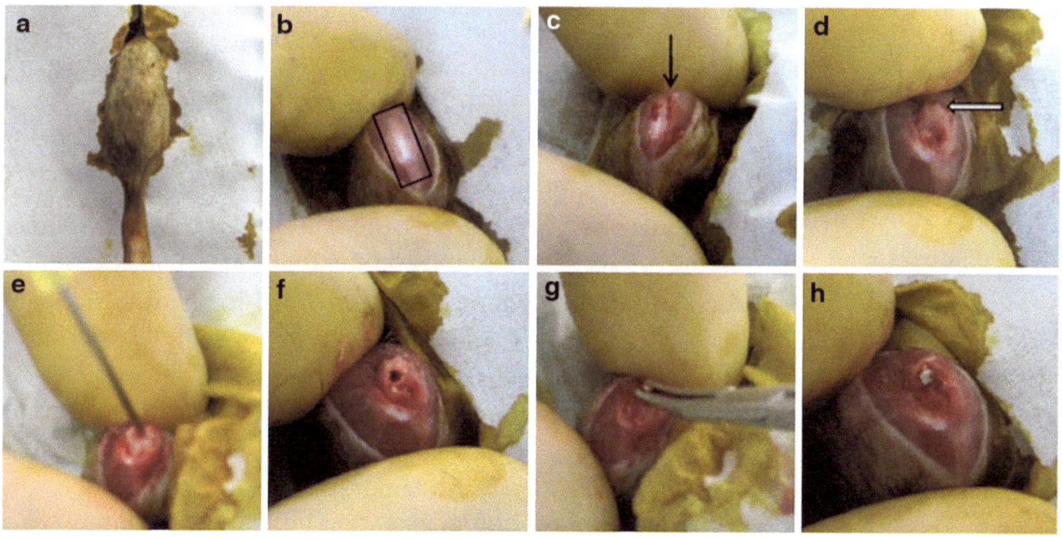

Fig. 1 Schematic of mouse arthroplasty surgery. (**a**) After removing hair and applying povidone-iodine solution, a sterile isolation drape is placed over the lower right leg. (**b**) An incision is made over the midline of the knee, through the skin and subcutaneous tissue, stopping at the depth of the fascia overlying the patella. At this point the patellar tendon is clearly visible (*box*). (**c**) An incision is made on the medial side of the patellar tendon (*arrow*). (**d**) Following retraction of the patella, the distal femoral trochlea is exposed (*hollow arrow*). (**e**, **f**) A needle is used to drill a hole through the trochlea into the intramedullary canal. (**g**, **h**) A piece of titanium wire is placed into the opening, extending approximately 1 mm into the patellar space

7. Identify the trochlea (Fig. 1d), then place a 23 gauge needle at the midpoint in the anterior/posterior plane, and twist gently back and forth by hand until the needle tip enters the intramedullary canal (*see* **Note 8**; Fig. 1e).
8. Continue to hand drill the femur with the needle until a scratch fit is obtained at the isthmus of the canal (Fig. 1f). This distance is usually 7–10 mm and sufficient to allow placement of the wire (*see* **Note 9**).
9. After preparing the canal as above, use the vascular surgery forceps to place the precut wire into the hole in the distal femur (Fig. 1g), leaving 1 mm of wire protruding into the knee joint (Fig. 1h).
10. Load the *S. epidermidis* suspension into a Hamilton syringe, and inoculate the protruding end of the wire with 2 μL of bacteria to achieve the desired infectious inoculum.
11. Close the arthrotomy with 1–2, figure of eight stitches, and then close the skin with a running stitch (*see* **Note 10**).
12. Immediately following wound closure, inject 100 μL of diluted buprenorphine subcutaneously near the surgical site for pain relief. Repeat every 8–12 h if signs of pain persist or as directed by your institution's IACUC.
13. Maintain animal's body temperature by placing cages under a heat lamp until they recover from anesthesia.

3.3 Sacrifice and Sample Collection

1. At the desired time points after infection, sacrifice animals by carbon dioxide inhalation (*see* **Note 11**) (Fig. 2).
2. Sterilize the lower right extremity with povidone-iodine cleansing solution (*see* **Note 12**).
3. Carefully dissect and remove the skin from the right extremity. Remove excess underlying adipose tissue and muscle.
4. Collect the tissue immediately proximal to the infection site, place in a 1.5 mL tube containing 500 μL of homogenization buffer, and weigh the sample.
5. Remove the wire and place it in a separate tube containing 1 mL of PBS.
6. Collect infected knee and femur, place in a separate tube containing 500 μL of homogenization buffer, and weigh.

3.4 Sample Processing and Analysis

1. Sonicate the wire for 5 min on ice to remove biofilm-associated bacteria. Perform serial dilutions of the effluent and plate on TSA to determine bacterial colonization.
2. Homogenize tissues using a Bullet Blender (10 min per sample). Afterward, being careful not to transfer any beads, pipette tissue homogenate into a clean 1.5 mL microcentrifuge tube. Perform serial dilutions and plate as above. Store the remaining

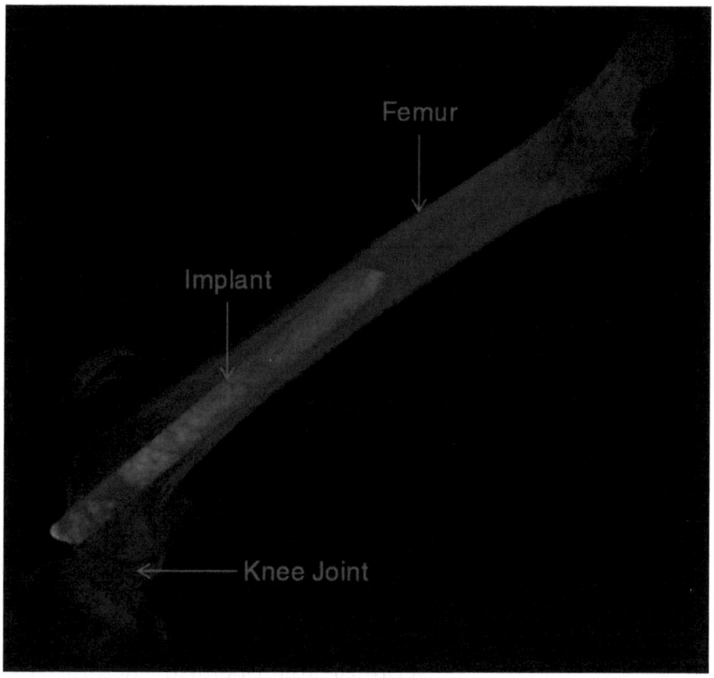

Fig. 2 X-ray microtomography (micro-CT) 3-D reconstruction illustrating the wire implant properly placed within the femoral medullary canal. Approximately 1 mm of wire tip is left protruding into the knee joint and serves as the site for bacterial inoculation

homogenate at −80 °C for later use to quantitate inflammatory mediators using microbead arrays or ELISAs (optional).

3. Homogenize the patella and femur using a handheld Polytron tissue homogenizer (30 s/sample) followed by processing in the Bullet Blender (10 min/sample). Pipette homogenates, sans beads, into clean 1.5 mL microcentrifuge tubes. Perform serial dilutions of each homogenate and plate on TSA to determine bacterial burdens. Store the remaining homogenate at −80 °C for later use to quantitate inflammatory mediators using microbead arrays or ELISAs (optional).
4. Express bacterial titers as Log10 CFU/mL for titanium wires or Log10 CFU/g wet tissue weight for knees, femurs, and surrounding tissues.
5. If desired, compare the expression of inflammatory mediators associated with biofilm-infected tissues using a mouse microbead array according to the manufacturer's instructions (MILLIPLEX; Millipore, Billerica, MA) or traditional ELISAs.
6. Analyze the microbead array results using a Bio-Plex workstation (Bio-Rad, Hercules, CA) or ELISA, and normalize values to the amount of total protein, quantified via a BCA protein assay, to correct for differences in tissue sampling size between animals.

4 Notes

1. Any sterile tubes with a capacity of at least 30 mL will work. We use 50 mL Nalgene Oak Ridge Centrifuge Tubes (Nalge Nunc International, Rochester, NY) with an SS-34 fixed angle rotor in a Sorvall RC 5C PLUS centrifuge (Thermo Fisher Scientific, Waltham, MA).
2. We have found that a 2 μL pipette with sterile tips may also be used.
3. The authors use SSB14B 0.9–2.0 mm stainless steel blend homogenization grinding beads for the tissues collected in this procedure.
4. Significant differences between experimental groups can be determined using a Student's *t*-test with Welch's correction for unequal variances (GraphPad Prism 4.03, GraphPad Software, Inc., La Jolla, CA). For all analyses, a *p*-value of less than 0.05 is considered statistically significant.
5. It is helpful to work under a well-lit laminar flow hood at a comfortable operating height. A sterile extremity drape is easily made by cutting a 5 × 5 in. square from the sterile towel or pad and cutting an approximately 1 in. slit in the center of the drape which the right leg of the mouse is placed through.
6. The authors have found it helpful to perform a quadriceps snip procedure by extending the arthrotomy incision into the quadriceps musculature proximally several millimeters in line with the muscle fibers. Distally, the incision is made to the level of the tibial tubercle. The usage of the quadriceps snip has greatly improved visualization of the knee joint.
7. The patella retraction must occasionally be done by feel when the patella is difficult to visualize due to size or blood in the operative field. When this is the case it remains easily palpable with the tip of the scalpel. To retract, the tip of the scalpel is slid medially and beneath the patella into the patellofemoral joint, and the patella is gently lifted and slid laterally over the edge of the lateral femur.
8. The tip of the needle should sit in the trochlea or slightly onto the medial condyle in line with the medullary canal. Be careful not to slide posterior between the condyles and into the popliteal fossa. By gently twisting the needle by hand, the distal femoral cortex is breached, and the needle will "find" the medullary canal of the femur. Aggressive drilling with the needle will often force the needle back out the lateral cortex or fracture the femur itself.
9. Drilling is not complete until the needle is entirely inserted into the canal.

10. If a quadriceps snip was performed, suturing of the underlying muscle and fascia should begin with a running stitch in the quadriceps before closing with figure-of-eight stitches over the patella.
11. Depending on your IACUC approval, this may also be performed by an overdose of isoflurane inhalation, followed by cervical dislocation as a secondary method to ensure death.
12. Alternatively, chlorhexidine may also be used for sterilization of the surgical field and instruments.

Acknowledgments

This work was supported by the NIH National Institute of Allergy and Infectious Disease (NIAID) P01 AI083211 to Tammy Kielian.

References

1. Fey PD, Olson ME (2010) Current concepts in biofilm formation of Staphylococcus epidermidis. Future Microbiol 5:917–933
2. McCann MT, Gilmore BF, Gorman SP (2008) Staphylococcus epidermidis device-related infections: pathogenesis and clinical management. J Pharm Pharmacol 60:1551–1571
3. Rohde H, Frankenberger S, Zahringer U et al (2010) Structure, function and contribution of polysaccharide intracellular adhesion (PIA) to Staphylococcus epidermidis biofilm formation and pathogenesis of biomaterial-associated infections. Eur J Cell Biol 89:103–111
4. Baillot R, Frechette E, Cloutier D et al (2012) Surgical site infections following transcatheter apical aortic valve implantation: incidence and management. Eur J Cardio Thorac Surg 7:122
5. Kane AD, Ndiaye MB, Pessinaba S et al (2012) Infections secondary to pacemaker implantation: a synopsis of six cases. Cardiovasc J Afr 23:e1–e4
6. Rosenthal ME, Dever LL, Moucha CS et al (2011) Molecular characterization of an early invasive Staphylococcus epidermidis prosthetic joint infection. Microb Drug Resist 17:345–350
7. Esteban J, Sandoval E, Cordero-Ampuero J et al (2012) Sonication of intramedullary nails: clinically-related infection and contamination. Open Orthop J 6:255–260
8. Sia IG, Berbari EF, Karchmer AW (2005) Prosthetic joint infections. Infect Dis Clin North Am 19:885–914
9. Zimmerli W, Trampuz A, Ochsner PE (2004) Prosthetic-joint infections. N Engl J Med 351:1645–1654
10. Anderl JN, Zahller J, Roe F et al (2003) Role of nutrient limitation and stationary-phase existence in Klebsiella pneumonia biofilm resistance to ampicillin and ciprofloxacin. Antimicrob Agents Chemother 47: 1251–1256
11. Ceri H, Olson ME, Stremick C et al (1999) The Calgary Biofilm Device: new technology for rapid determination of antibiotic susceptibilities of bacterial biofilms. J Clin Microbiol 37:1771–1776
12. Montanaro L, Speziale P, Campoccia D et al (2011) Scenery of Staphylococcus implant infections in orthopedics. Future Microbiol 6:1329–1349
13. Dominguez-Herrera J, Docobo-Perez F, Lopez-Rojas R et al (2011) Efficacy of daptomycin versus vancomycin in an experimental model of foreign-body and systemic infection caused by biofilm producers and methicillin-resistant staphylococcus epidermidis. Antimicrob Agents Chemother 56: 613–617
14. Del Pozo JL, Rouse MS, Euba G et al (2009) The electricidal effect is active in an experimental model of staphylococcus epidermidis chronic foreign body osteomyelitis. Antimicrob Agents Chemother 53: 4064–4068
15. Ribeiro M, Monteiro F, Ferraz MP (2012) Infection of orthopedic implants with emphasis on bacterial adhesion process and techniques used in studying bacterial-material interactions. Biomatter 2:176–194

Chapter 17

A Mouse Model of *Staphylococcus* Catheter-Associated Biofilm Infection

Cortney E. Heim, Mark L. Hanke, and Tammy Kielian

Abstract

Biofilms are adherent communities of bacteria contained within a complex matrix. Staphylococcal species are frequent etiological agents of device-associated biofilm infections in humans that are highly recalcitrant to antimicrobial therapy and alter host immune responses to facilitate bacterial persistence. Here we describe a mouse model of catheter-associated biofilm infection, which can be utilized to investigate the importance of various staphylococcal determinants on disease progression as well as the host immune response to staphylococcal biofilms.

Key words *Staphylococcus aureus*, *Staphylococcus epidermidis*, Biofilm, Catheter

1 Introduction

Biofilms are adherent communities of bacteria contained within a complex matrix. From a clinical standpoint, biofilm infections of native tissues or medical devices represent a serious therapeutic challenge, since organisms are typically recalcitrant to conventional antibiotics [1, 2]. Medical device-related infections are typified by high morbidity, with their clinical management often requiring device removal [3–6]. In situations where salvaging infected hardware is attempted, the failure rate is often high, despite prolonged antimicrobial therapy [7–13]. To date, staphylococcal species remain one of the major causes of both health care-associated (HA) as well as community-associated (CA) infections. *Staphylococcus epidermidis* (*S. epidermidis*) is a frequent etiological agent of biofilm infections on medical devices, including indwelling catheters and prostheses [14–16], whereas *Staphylococcus aureus* (*S. aureus*) biofilms are commonly associated with tissue infections, such as endocarditis [17] and osteomyelitis [18], although this organism is also a major cause of device-associated infections [19, 20]. Based on their chronicity, debilitating nature, and economic impact, biofilm infections are of paramount significance in modern medicine.

Paul D. Fey (ed.), *Staphylococcus Epidermidis: Methods and Protocols*, Methods in Molecular Biology, vol. 1106,
DOI 10.1007/978-1-62703-736-5_17, © Springer Science+Business Media, LLC 2014

Therefore, it is imperative to better decipher the interactions between staphylococcal biofilms and the host to devise novel treatment strategies to combat these devastating infections.

Here we describe a model of subcutaneous catheter-associated biofilm infection that has been utilized by a number of laboratories [21, 22]. This model allows for the investigation of host–pathogen interactions at the site of a chronic infection. Advantages of the model are its easy tissue accessibility for monitoring infection, straightforward device placement, and relatively high-throughput nature compared to other models that require more laborious surgery. However, it should be noted that this is a static model, which does not recapitulate the sheer forces that a biofilm would encounter in an indwelling intravascular catheter, since the device is placed subcutaneously in the flank. Nonetheless, this model provides an excellent tool to investigate various aspects of host–pathogen biofilm interactions and can be used with bacterial mutant strains or mice deficient for various immune-related molecules to identify critical disease determinants.

2 Materials

Prepare all solutions using sterile ultrapure water or PBS and analytical grade reagents. Prepare and store all reagents at room temperature (unless indicated otherwise). Strict adherence should be made to follow all waste disposal regulations when handling infectious hazardous waste materials.

2.1 Bacterial Preparation

1. Brain Heart Infusion (BHI): Suspend 37 g of powder in 1 L of purified water. Mix thoroughly. Autoclave at 121 °C for 15 min.
2. One liter flask(s).
3. Centrifuge and appropriate tubes.
4. Sterile saline or PBS.
5. Tubes for dilution plating.
6. TSA plates supplemented with 5 % sheep blood.

2.2 Catheter Insertion and Inoculation

1. Mouse strain: can be any strain desired, including outbred and inbred strains. Our laboratory typically uses C57BL/6 mice, since the majority of mouse knockout strains have been generated on the C57BL/6 background. Animals typically between 8 and 10 weeks of age are used (*see* **Note 1**).
2. Small animal scale.
3. Mouse ear punch tool.
4. Anesthesia: 2,2,2-Tribromoethanol (TBE). Prepare a stock solution of TBE by mixing 25 g of TBE with 15.5 mL of tert-amyl

alcohol in a dark bottle on a stir plate protected from light (will take 12–24 h at room temperature until the TBE is completely dissolved). Once completely dissolved, wrap the TBE stock solution in foil and keep at room temperature (stock solution is both hydroscopic and photosensitive). Prepare a working solution of TBE prior to surgery by mixing 1 mL of TBE stock with 39 mL PBS and stir protected from light until completely dissolved. Filter-sterilize the working solution and store in the dark at 4 °C. The working solution, stored properly, can be used for 2 weeks. As an alternative to TBE, ketamine–xylazine or isoflurane inhalation anesthesia can be used.

5. Electric Clippers to remove fur.
6. Puralube® vet ointment or comparable ophthalmic ointment.
7. Povidone–Iodine Prep Pads.
8. Surgical Tools: scissors, blunt probe, forceps.
9. Sterile 14-gauge teflon intravenous catheter 1 cm in length (Exel International, St. Petersburg, FL).
10. Vetbond Tissue Adhesive (3 M, St. Paul, MN) for closing the skin surgical incision site.

2.3 Sacrifice and Sample Collection

1. Dissection tools including: scissors, forceps.
2. Sterile PBS.
3. 5 mL polystyrene tubes.
4. Homogenization buffer: PBS supplemented with protease inhibitor cocktail tablet (Roche Diagnostics, Indianapolis, IN). Store working aliquots at 4 °C or −20 °C for long term storage.
5. 1.5 mL or 2 mL microcentrifuge tubes.
6. Small animal scale.

2.4 Sample Processing and Analysis

1. 1.5 mL microcentrifuge tubes.
2. Bullet Blender (Next Advance Inc., Averill Park, NY) and grinding beads (*see* **Note 2**).
3. TSA plates supplemented with 5 % sheep blood.
4. Graphing and statistical software.
5. Mouse microbead array (MILLIPLEX; Millipore, Billerica, MA).
6. Bio-Plex workstation (Bio-Rad, Hercules, CA).
7. BCA protein assay (Bio-Rad, Hercules, CA).

2.5 Scanning Electron Microscopy

1. 0.1 M Sorensen's phosphate buffer containing 2 % glutaraldehyde and 2 % paraformaldehyde.
2. Ethanol.
3. Pelco CPD2 critical point dryer (Ted Pella, Redding, CA).

4. Aluminum stubs with carbon tabs.
5. Colloidal silver.
6. Gold–Palladium.
7. Hummer VI sputter coater (Anatech, Battle Creek, MI).
8. Quanta 200 scanning electron microscope (FEI, Hillsboro, OR) operated at 25 kV.

3 Methods

3.1 Bacterial Preparation

1. Inoculate 50 mL of bacterial growth medium in a 500 mL baffled flask. Incubate at 37 °C with aeration (200 rpm) for 12–18 h.
2. The following day, spin down 1 mL of culture at 20,000 × *g* for 5 min. Decant supernatant and wash cell pellet with 1 mL sterile saline or PBS. Spin as before. Decant the supernatant and resuspend cell pellet with 1 mL sterile saline or PBS.
3. Read optical density using photospectrometer at 620 nm. Estimate bacterial titer and perform serial dilutions to achieve desired inoculum of 10^3 CFU/20 μL in PBS.

3.2 Catheter Insertion and Inoculation

1. Weigh mouse, anesthetize by intraperitoneal injection of TBE (15 μL TBE working solution/gram of weight) or alternative anesthetic, and ear punch for animal identification.
2. Remove hair from the left flank region using electric hair clippers. Apply ophthalmic ointment to both eyes to prevent them from drying out during the surgical procedure.
3. Sterilize the surgical field (i.e., flank) with povidone–iodine prep pads
4. Make a small s.c. incision in the flank just above the hind leg using a sterile blade
5. Insert a blunt probe to create a s.c. pocket for advancing the catheter. Note that catheters must be advanced a sufficient distance from the primary incision site to avoid catheter extrusion from the mouse over time.
6. Using forceps insert the sterile catheter (*see* **Note 3**)
7. Seal the skin incision with Vetbond tissue adhesive
8. Inject desired bacterial inoculum (i.e., we utilize 1,000 CFU USA300 LAC::*lux* to visualize bacterial burdens using an *In Vivo Imaging System*; IVIS) in 20 μL of sterile PBS through the skin into the catheter lumen, using a 3/10 cc, 29G × ½ ″ syringe.

3.3 Monitoring Infection

1. The extent of biofilm formation can be monitored longitudinally by IVIS while animals are subjected to isoflurane inhalation anesthesia (Fig. 1) (*see* **Note 4**)

3.4 Sacrifice

1. At day 3, 7, or 14 post-infection, sacrifice animals with an overdose of inhaled isoflurane or alternative AVMA-approved procedure.
2. Sterilize left flank with povidone–iodine prep pad. This is important to prevent potential sample contamination with skin microflora (a particular concern with *S. epidermidis*). This can be assessed by examining an additional cohort of mice implanted with sterile catheters, where no bacterial growth should be observed.
3. Remove the section of flank containing the catheter and associated tissue. Note that the longer infected catheters remain in mice the more fibrotic tissue is detected around the device. Catheters will not always be immediately visible upon excision until this surrounding host tissue is removed and collected.
4. Separate catheter from surrounding host tissue and place in 1 mL PBS on ice for sonication to quantitate bacterial burdens on the device.

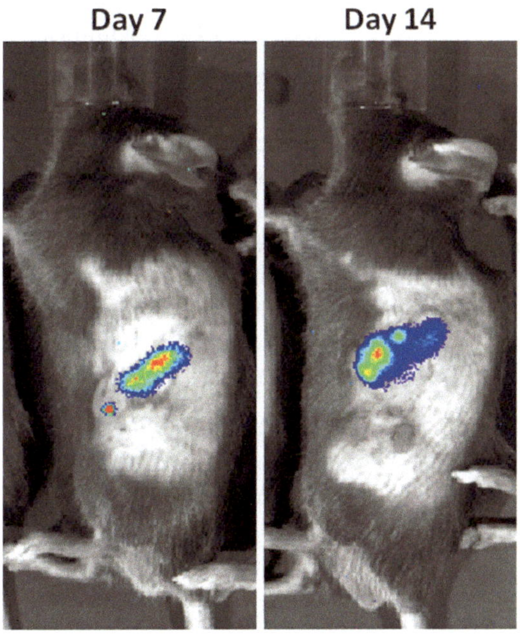

Fig. 1 Visualization of *S. aureus* biofilm infection using IVIS. WT mice were infected with USA300 LAC::*lux* in the lumen of surgically implanted catheters to establish biofilm infection. At the indicated time points post-infection, mice were subjected to IVIS imaging to visualize the extent of biofilm infection. Images presented are of the same animal at days 7 and 14 post-infection

5. Collect tissue surrounding catheter, weigh, and place in 500 μL homogenization buffer on ice for determining bacterial burdens and production of inflammatory mediators, if applicable.
6. To evaluate the possibility of bacterial dissemination, the kidney and heart can also be collected in 500 μL PBS and weighed to prepare homogenates for bacterial titer determination.

3.5 Sample Processing and Analysis

1. Sonicate catheters for 5 min on ice. Perform serial dilutions of the effluent and plate on TSA supplemented with 5 % sheep blood to determine bacterial colonization.
2. Homogenize catheter-associated tissue and any organs collected using a Bullet Blender (10 min per sample). Pipette homogenates, sans beads, into clean 1.5 mL microcentrifuge tubes. Perform serial dilutions of each homogenate and plate on TSA to determine bacterial burdens. Store remaining homogenates at −80 °C for later use to quantitate inflammatory mediators using microbead arrays or ELISAs (optional).
3. Express bacterial titers as Log10 CFU/mL for catheters or Log10 CFU/g wet tissue weight for catheter associated tissues and other organs.
4. If desired, compare the expression of inflammatory mediators associated with biofilm-infected tissues using a mouse microbead array according to the manufacturer's instructions (MILLIPLEX; Millipore, Billerica, MA) or traditional ELISAs.
5. Analyze the microbead array results using a Bio-Plex workstation or ELISA and normalize values to the amount of total protein, quantified via a BCA protein assay, to correct for differences in tissue sampling size between animals.

3.6 Scanning Electron Microscopy

1. If the ultrastructural characteristics of in vivo biofilms will be investigated, the catheter and surrounding tissue can be harvested as a single unit from mice and fixed with 2.5 % glutaraldehyde in 0.1 M PBS (pH 7.4) for 10 min
2. Cut specimens longitudinally with a razor blade, continue fixing for no more than 50 min
3. Wash samples thoroughly in three changes 0.1 M PBS for 15 min each.
4. Perform secondary fixation by placing samples in 1 % osmium tetroxide in 0.1 M PBS for 60 min.
5. Wash thoroughly in three changes 0.1 M PBS for 15 min each
6. Dehydrate samples using a graded series of alcohol washes for 10 min each
   (a) 25 % EtOH
   (b) 50 % EtOH

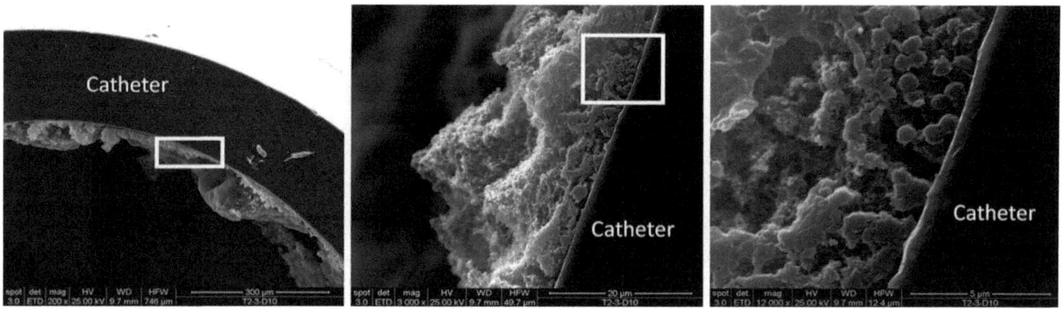

Fig. 2 Catheter-associated biofilm growth in vivo. Catheters were isolated from mice at day 10 following *S. aureus* infection and processed for scanning electron microscopy analysis. The smooth surface at the periphery of the images represents the catheter with biofilm visible on the internal face. *Left*: Original magnification ×200 demonstrating the irregular undulating pattern of the biofilm surface. *Center*: Higher magnification of a tower from the image shown at *left*, revealing a predominantly hollow interior with numerous cocci at the margins (original magnification ×800). *Right*: Original magnification ×3,000 of an *S. aureus* cluster within the tower depicted in the center panel

(c) 70–75 % EtOH

(d) 90–95 % EtOH

(e) 100 % EtOH × 2

7. Critical point dry tissue using a Pelco CPD2 critical point dryer (Ted Pella, Redding, CA).
8. Dried specimens are mounted onto aluminum stubs with carbon tabs and colloidal silver paste and sputter coated with gold–palladium using a Hummer VI sputter coater.
9. View samples using Quanta 200 scanning electron microscope or equivalent operated at 25 kV (Fig. 2).

3.7 Statistical Analysis

1. Significant differences between experimental groups can be determined using a Student's t-test with Welch's correction for unequal variances. For all analyses, a p-value of less than 0.05 is considered statistically significant.

4 Notes

1. All animal procedures should be carried out in accordance with the National Institutes of Health Guidelines for the Care and Use of Laboratory Animals and be approved by institutional Animal Care and Use Committees.
2. The authors use SSB14B 0.9–2.0 mm stainless steel blend homogenization grinding beads for the tissue collected in this procedure.
3. Coating of artificial surfaces and implanted medical devices with platelets and host ECM proteins, such as fibronectin,

facilitate *S. aureus* adherence in vitro [23]. Likewise, others have reported enhanced adhesion with various plasma proteins, including fibronectin, with coagulase-negative staphylococci [24]. These interactions are facilitated by the numerous microbial surface components recognizing adhesive matrix molecules (MSCRAMMs) that are expressed by staphylococci, which have known binding affinity for monomeric collagen and fibronectin [25]. However, others have demonstrated that fibronectin and its proteolytic fragments inhibited *S. epidermidis* adhesion to plastic surfaces [26]. This might be explained by either the growth phase when bacteria were harvested (since MSCRAMMs are maximally expressed during log phase growth in broth cultures) or the fact that *S. epidermidis* is more adept at binding to inert surfaces compared to *S. aureus*. In this case, coating with exogenous molecules is more dispensable and conceivably, could deter *S. epidermidis* binding due to charge repulsion, although this remains speculative.

4. *I*n *V*ivo *I*maging *S*ystems (IVIS) have been used to monitor the course of biofilm development in several animal models, making it an attractive tool to evaluate how the host immune response impacts biofilm growth. Bioluminescence imaging allows for the direct visualization of bacterial growth or mammalian gene expression, utilizing reporter mouse strains engineered to express luciferase under the control of promoters pivotal in host immunity to gram-positive bacteria (i.e., iNOS, TLR2, and NF-κB), using reporter constructs where the luciferase gene has been stably expressed to monitor expression throughout biofilm development [26]. This technology takes advantage of photon emission from either the mammalian or bacterial luciferase genes (*luc* and *lux*, respectively). The bacterial *lux* gene encodes an enzyme that constitutively emits photons, whereas the mammalian enzyme requires a luciferin substrate be injected into animals immediately prior to imaging since an endogenous substrate is not available. The IVIS system provides an innovative approach to evaluate regional expression patterns of immune genes, using luciferase-reporter mice, or to evaluate the kinetics of *Staphylococcus* biofilm growth longitudinally in the same animal. Another benefit in the use of IVIS is the ability to image the same cohort of animals over an entire experiment, providing a longitudinal assessment of bacterial growth or inflammatory gene expression. This information is not attainable with standard quantitative cultures or histology, since mice must be sacrificed at specified time points to evaluate these parameters.

Acknowledgments

This work was supported by the NIH National Institute of Allergy and Infectious Disease (NIAID) P01 AI083211 Project 4 to T.K.

References

1. Donlan RM, Costerton JW (2002) Biofilms: survival mechanisms of clinically relevant microorganisms. Clin Microbiol Rev 15:167–193
2. Stewart PS, Costerton JW (2001) Antibiotic resistance of bacteria in biofilms. Lancet 258: 135–138
3. Lew DP, Waldvogel FA (1997) Osteomyelitis. N Engl J Med 326:999–1007
4. NIH Concenus Conference (1995) Total hip replacement. JAMA 273:1950–1956
5. Garvin KL (1995) Infection after total hip arthroplasty. J Bone Joint Surg Am 77:1576–1588
6. Morscher E, Herzog R, Bapst R et al (1995) Management of infected hip arthroplasty. Orthop Int 3:343–351
7. Schoifet SD, Morrey BF (1990) Treatment of infection after total knee arthroplasty by debridement with retention of the components. J Bone Joint Surg Am 72:1383–1390
8. Rasul AT, Tsukayama D, Gustilo RB (1991) Effect of time on onset and depth of infection on the outcome of total knee arthroplasty infections. Clin Orthop 273:98–103
9. Burger RR, Basch T, Hopson CN (1991) Implant salvage in infected total knee arthroplasty. Clin Orthop 273:105–111
10. Hartman MB, Fehring TK, Jordan L et al (1991) Periprosthetic knee sepsis: the role of irrigation and debridement. Clin Orthop 273:113–118
11. Tsukayama DT, Gustilo RB (1991) Suppressive antibiotic therapy in chronic prosthetic joint infections. Orthopedics 14:841–844
12. Wilson MG, Kelley K, Thornhill TS (1990) Infection as a complication of total knee-replacement arthroplasty. J Bone Joint Surg Am 72:878–883
13. Brandt CM, Sistrunk WW, Duffy MC et al (1997) *Staphylococcus aureus* prosthetic infection treated with debridement and prosthesis retention. Clin Infect Dis 24:914–919
14. Fitzpatrick F, Humphreys H, O'Gara JP (2005) The genetics of staphylococcal biofilm formation—will a greater understanding of pathogenesis lead to better management of device-related infection? Clin Microbiol Infect 11:967–973
15. Otto M (2008) Staphylococcal biofilms. Curr Top Microbiol Immunol 322:207–228
16. Fey PD (2010) Modality of bacterial growth presents unique targets: how do we treat biofilm-mediated infections? Curr Opin Microbiol 13:610–615
17. Fitzsimmons K, Bamber AL, Smalley HB (2010) Infective endocarditis: changing aetiology of disease. Br J Biomed Sci 67:35–41
18. Zuluaga AF, Galvis W, Saldarriaga JG et al (2006) Etiologic diagnosis of chronic osteomyelitis: a prospective study. Arch Intern Med 166:95–100
19. Donlan RM (2001) Biofilms and device-associated infections. Emerg Infect Dis 7: 277–281
20. Darouiche RO (2004) Treatment of infections associated with surgical implants. N Engl J Med 350:1422–1429
21. Rupp ME, Ulphani JS, Fey PD et al (1999) Characterization of the importance of polysaccharide intercellular adhesin/hemagglutinin of *Staphylococcus epidermidis* in the pathogenesis of biomaterial-based infection in a mouse foreign body infection model. Infect Immun 67:2627–2632
22. Beenken KE, Spencer H, Griffin LM et al (2012) Impact of extracellular nuclease production on the biofilm phenotype of *Staphylococcus aureus* under in vitro and in vivo conditions. Infect Immun 80:1634–1638
23. Foster TJ (1996) Staphylococcus. In: Baron S (ed) Medical microbiology. University of Texas Medical Branch at Galveston, Galveston, TX
24. Herrmann M, Vaudaux PE, Pittet D et al (1988) Fibronectin, fibrinogen, and laminin act as mediators of adherence of clinical staphylococcal isolates to foreign material. J Infect Dis 158:693–701
25. Rivera J, Vannakambadi G, Höök M et al (2007) Fibrinogen-binding proteins of Gram-positive bacteria. Thromb Haemost 98:503–511
26. Dunne WM, Burd EM (1993) Fibronectin and proteolytic fragments of fibronectin interfere with the adhesion of Staphylococcus epidermidis to plastic. J Appl Bacteriol 74:411–416

Chapter 18

Generation of a Central Nervous System Catheter-Associated Infection in Mice with *Staphylococcus epidermidis*

Jessica N. Snowden

Abstract

Animal models are valuable tools for investigating the in vivo pathogenesis of *Staphylococcus epidermidis* infections. Here, we present the procedure for generating a central nervous system catheter-associated infection in a mouse, to model the central nervous system shunt infections that frequently complicate the treatment of hydrocephalus in humans. This model uses stereotactic guidance to place silicone catheters, pre-coated with *S. epidermidis*, into the lateral ventricles of mice. This results in a catheter-associated infection in the brain, with concomitant illness and inflammation. This animal model is a valuable tool for evaluating the pathogenesis of bacterial infection in the central nervous system, the immune response to these infections and potential treatment options.

Key words Central nervous system, Catheter infection, Mouse model, Shunt infection

1 Introduction

Cerebrospinal fluid (CSF) shunt infections are a frequent and serious complication in the treatment of hydrocephalus in the pediatric population [1]. The most common organism responsible for these infections, *Staphylococcus epidermidis*, is known to form persistent catheter-associated infections [2, 3]. These infections are recalcitrant to systemic antibiotic therapy and make it difficult to manage shunt infections nonsurgically, such that catheter removal is currently required to effectively treat these infections. To better understand the pathogenesis and immune response to these central nervous system catheter infections, a model has been developed that generates a catheter-associated infection with *S. epidermidis* in the mouse. This technique results in a consistent catheter-associated infection with *S. epidermidis*, by coating silicone catheters with *S. epidermidis* and inserting the catheter into the lateral ventricle of the mouse. Establishing a catheter model within the central nervous

Paul D. Fey (ed.), *Staphylococcus Epidermidis: Methods and Protocols*, Methods in Molecular Biology, vol. 1106,
DOI 10.1007/978-1-62703-736-5_18, © Springer Science+Business Media, LLC 2014

system was necessary because the immune response in this compartment often differs from that seen in the periphery [4]. Mice implanted with infected catheters using this technique demonstrate elevated pro-inflammatory cytokine and chemokine levels compared to sterile catheters. In addition, the majority of bacteria are associated with the catheter surface with minimal spread to surrounding tissues. Based on these findings, we propose that this model serves as a powerful tool to identify important factors in the pathogenesis of central nervous system catheter infections.

2 Materials

1. Male C57BL/6 mice at 8 weeks of age (*see* **Notes 1** and **2**).
2. *S. epidermidis* log-phase bacterial solution at $1\text{–}2 \times 10^{10}$ colony forming units (cfu) per ml brain-heart infusion (BHI) broth.
3. Hollow bore silicone catheters (2 mm length, 1 mm diameter; Cook Medical Inc, Bloomington IN) (*see* **Note 3**).
4. Ketamine–xylazine preparation: Mix 1 ml of ketamine (concentration 100 mg/ml), 0.5 ml xylazine (concentration 20 mg/ml), and 8.5 ml of PBS or normal saline 0.9 %; administer via 1 ml syringe with 23–25 gauge 5/8 in. needle (*see* **Note 4**).
5. Aseptic surgical supplies including: razor blade; straight, pointed forceps or tweezers; curved forceps; 18 gauge needle; sterile gauze.
6. Rodent stereotaxic apparatus equipped with a Cunningham mouse adaptor (Stoelting, Wood Dale IL).
7. Vetbond surgical glue (3 M, St. Paul MN).
8. Lubricating eye ointment, heat lamps for postoperative care.

3 Methods

All surgical procedures are carried out under aseptic conditions, using aseptic catheters and surgical instruments.

3.1 Preparation of Infected Catheters

2 mm sections of hollow bore silicone catheter are placed into a 500 μl capped tube with 300 μl of *S. epidermidis* bacterial broth (concentration $1\text{–}2 \times 10^{10}$ cfu/ml). Briefly vortex the tube (5 s) to ensure even mixing and coating of the catheter and check that the catheter is completely submerged in the bacterial broth. This tube is then incubated (stationary) at 37 °C for 24 h. After 24 h of incubation, place each tube on ice for transport to the site of animal surgery (*see* **Notes 5** and **6**).

3.2 Implantation of Infected Catheters

Male C57BL/6 mice at 8 weeks of age are utilized in this model.

1. The mouse should be weighed (grams) and marked with ear punch for tracking of his clinical course.
2. Each mouse then receives an intraperitoneal injection of ketamine–xylazine (0.1 ml per 10 g body weight i.p.) to induce general anesthesia. After anesthesia, carefully shave the dorsal surface of the head of each animal to clear the surgical field of fur (*see* **Note 7**).
3. The surgical site (dorsal surface of the head) is then scrubbed with betadine to reduce the surface bacterial burden and to remove any loose fur.
4. Using a sterile razor blade, make a 1-cm longitudinal incision in the scalp to expose the underlying skull sutures.
5. Position the mouse in a rodent stereotaxic apparatus equipped with a Cunningham mouse adaptor (Stoelting, Wood Dale IL) that contains highly precise rulers (in mm) for accurate positioning in the x-, y-, and z-planes. This allows for accurate catheter placement into the left ventricle using the following coordinates relative to bregma: +0.02 mm rostral, +1.0 mm lateral, and −2.0 mm deep from the surface of the brain (Fig. 1) [5] (*see* **Note 8**).
6. Using an 18 gauge hollow bore needle, carefully make a small burr hole at these coordinates using a gentle twisting motion vertically to perforate the skull, and then at a 45° angle in relation to the skull surface to carefully widen the burr hole (*see* **Note 9**).
7. Remove an infected catheter from the bacterial broth and gently insert it vertically into the burr hole using straight, pointed forceps or tweezers. The catheter should be inserted until the outer edge of the catheter is even with the level of the skull.
8. Following catheter placement, any blood or fluid should be absorbed using sterile gauze to minimize irritation. The burr hole is carefully sealed with bone wax and the skin incision closed using Vetbond surgical glue (3 M, St. Paul, MN) (*see* **Note 10**). Animals also receive eye ointment (Lacrilube, Duratears) to prevent the eyes from drying out during anesthesia due to loss of the palpebral reflex.
9. During the postoperative period, animals should be placed under a heat lamp to prevent hypothermia (*see* **Note 11**).

4 Notes

1. Obtain approval from your local Institutional Animal Care and Use Committee prior to performing any experiments utilizing mice.

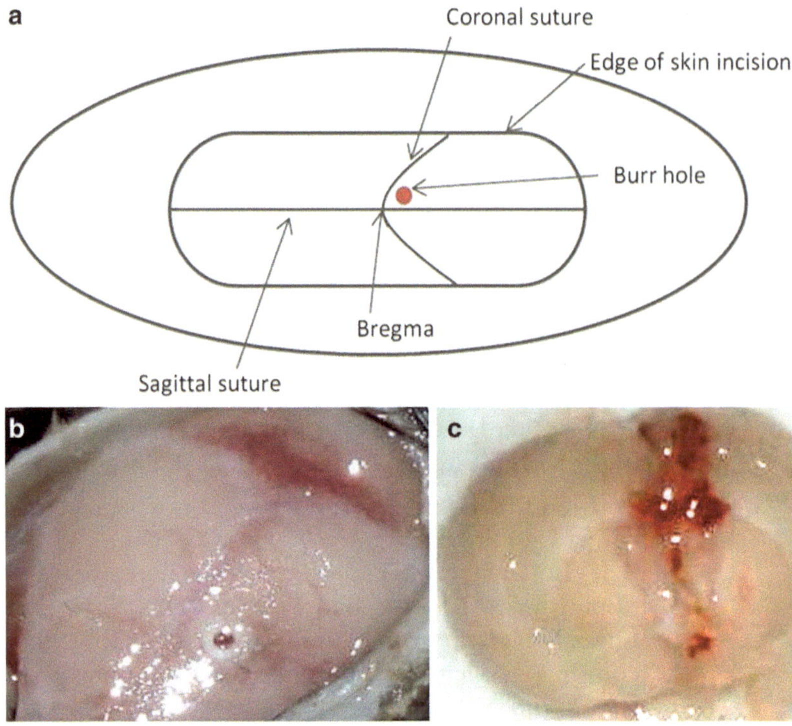

Fig. 1 Overview of implantation procedure. Silicone catheters are incubated with 1×10^{10} cfu/ml *S. epidermidis* for 24 h to facilitate bacterial attachment. Catheter fragments are stereotactically inserted into the lateral ventricle of C57BL/6 mice. This location (Images a and b, from bregma: +0.02 mm rostral, +1 mm lateral, −2 mm deep) was chosen because the lateral ventricles are fairly large in diameter and more easily accessible at this location (Image c). Copyright © American Society for Microbiology, Infection and Immunity, 2012. 80: 3206–3214. doi: 10.1128/IAI.00645-12

2. Young adult C57BL/6 mice (6–8 weeks of age) were utilized for these studies since our long-term goal is to eventually utilize various genetically engineered knockout or transgenic mouse strains for candidate immune molecules that are bred on a C57BL/6 background. This age range facilitates the ease of catheter placement within the ventricle due to the larger size of the brain compared to younger animals. In addition, older and slightly larger animals will have a lower mortality rate, which can be a complication in younger animals. These studies were performed using female mice in early trials with *S. aureus*, but mortality rates were much higher. This is most likely due to the smaller size of female mice and their inability to survive the weight loss associated with their illness, rather than an inherent immune susceptibility as similar mortality rates are seen with smaller male mice. This procedure could likely be performed on other strains or ages of mice, although this has not been specifically investigated.

3. For these studies, we utilized 2 mm sections of pediatric intravenous catheters manufactured by Cook Medical Inc.
4. The combination of ketamine and xylazine is often used in surgical procedures involving rodents as ketamine is an effective dissociative agent and xylazine a powerful sedative and analgesic. This mixture provides a surgical level of anesthesia for 15–30 min in most mice and sedation for 1–2 h following initial injection. The mixture loses potency over time, so we would recommend preparing a fresh mixture on the day of the planned surgical procedure [6].
5. In establishing this mouse model, we utilized *S. epidermidis* strain 1457, a well-characterized clinical strain that has been utilized in multiple animal models of foreign body infection [7–9].
6. This procedure has also been performed using *S. aureus* and with sterile catheters (incubated with PBS or mouse serum), using an identical surgical approach [4]. For infection with *S. aureus*, the catheter preparation has minor differences in dose (2×10^4 cfu/ml bacterial broth), incubation time (4 h) and pre-coating of the catheters with mouse serum.
7. Note that since ketamine is a controlled substance, a detailed log of its use should be maintained in addition to securing the drug in a lock box. Anesthetic depth should be monitored in animals by evaluating the pedal withdrawal and tail pinch reflexes in addition to respiration rate and depth. Any mice reacting to painful stimuli should receive a second injection of ketamine–xylazine, at half the original dose, or ketamine alone to ensure that a surgical plane of anesthesia has been achieved prior to initiation of the procedure. Care must be taken with repeated administration of the ketamine–xylazine mixture as xylazine has a longer duration of action than ketamine, increasing the risk of overdose with repeated administrations [6].
8. These coordinates were selected for several reasons: first, at this anatomical location, the lateral ventricles are fairly large in diameter and more easily accessible and second, our extensive work with a mouse experimental brain abscess model has established that mortality rates are reduced with more rostral infection placements due to less pressure on critical brainstem centers from the edematous response that ensues following infection. This also replicates the site of catheter placement in children.
9. Using a marker to note the site of the coordinates and then removing the mouse from the stereotactic apparatus for the burr hole and catheter implantation may be technically easier for some.

10. Glue is generally less irritating to mice than sutures or staples. In addition, we did not want to introduce another foreign body to potentiate spread of infection from the catheter site.
11. In monitoring the mice after the procedure, we have observed that they experience weight loss (up to 10–15 % initial body weight) and signs of illness, including decreased mobility and ruffled fur, in the first 3–5 days following the implantation of an infected catheter. The mortality rate is very low (<5 %) and most mice return to baseline weight and behavior following the first week.

Acknowledgements

This work is supported by NIH grant 1K08NS069812-01 and the Cheryl Ann Lozier Memorial Research Fund.

References

1. McGirt MJ, Zaas A, Fuchs HE et al (2003) Risk factors for pediatric ventriculoperitoneal shunt infections and predictors of infectious pathogens. Clin Infect Dis 36:858–62
2. Kockro RA, Hampl JA, Jansen B et al (2000) Use of scanning electron microscopy to investigate the prophylactic efficacy of rifampin-impregnated CSF shunt catheters. J Med Microbiol 49:441–450
3. Fux CA, Quigley M, Worel AM et al (2006) Biofilm-related infections of cerebrospinal fluid shunts. Clin Microbiol Infect 12:331–337
4. Snowden J, Beaver M, Smeltzer M et al (2012) Biofilm infected intracerebroventricular shunts elicit inflammation within the central nervous system. Infect Immun 80:3206–3214
5. Paxinos G, Franklin K (2012) The mouse brain in stereotaxic coordinates. Academic, Waltham, MA
6. Ketamine-xylazine combination for rodent anesthesia. National Cancer Institute Laboratory Animal Sciences Program. http://ncifrederick.cancer.gov/rtp/lasp/intra/acuc/beth/Ketamine Xylazine.asp. Accessed 28 Dec 2012
7. Shahrooeil M, Hira V, Stijlemans B et al (2009) Inhibition of *Staphylococcus epidermidis* biofilm formation by rabbit polyclonal antibodies against the SesC protein. Infect Immun 77:3670–3678
8. Olsen ME, Slater SR, Rupp ME et al (2010) Rifampicin enhances activity of daptomycin and vancomycin against both a polysaccharide intercellular adhesion (PIA)-dependent and -independent *Staphylococcus epidermidis* biofilm. J Antimicrob Chemother 65:2164–2171
9. Kronfrost KD, Mancuso CJ, Pettengill M et al (2012) A neonatal model of intravenous *Staphylococcus epidermidis* infection in mice, 24 h old, enables characterization of early innate immune responses. PLoS One 7:e43897

Chapter 19

Rat Jugular Catheter Model of Biofilm-Mediated Infection

Carolyn R. Schaeffer, Keith M. Woods, and G. Matthew Longo

Abstract

Staphylococcus epidermidis is now recognized as the primary cause of nosocomial catheter-mediated infections. Bacteria may be introduced exogenously via contamination of the catheter hub or insertion site and endogenously from sepsis. The in vivo model described in this chapter examines the infection resulting from hematogenous seeding of jugular vein catheters.

Key words *Staphylococcus epidermidis*, Central venous catheter (CVC), Jugular vein, Biofilm, Cyclophosphamide (CP), Rat model

1 Introduction

Staphylococcus epidermidis is the organism most frequently isolated from infected intravenous catheters, in particular central venous catheters (CVC) [1]. These infections are biofilm-mediated and as a consequence are recalcitrant to antibiotics and able to evade host immune defenses, ultimately necessitating device removal [2]. The dynamic nature of the infective process cannot be accurately modeled in vivo. For this reason, representative animal models are essential for ascertaining the mechanisms leading to disease, as well as improving patient treatment and outcomes.

Here we describe a model of *S. epidermidis* CVC biofilm infection resulting from hematogenous device colonization. The literature contains multiple variations of the rat catheter model originally published by Rupp et al. [3–6]. The following method isolates the process of bacterial adherence and attachment, while alleviating the need for cumbersome jackets or restraints, and decreasing the likelihood of exogenous contamination. Additionally, implantation of the catheter prior to bacterial inoculation allows a fibrin sheath to begin forming. Coating of foreign materials with host proteins models the clinical situation and is an important component of

Paul D. Fey (ed.), *Staphylococcus Epidermidis: Methods and Protocols*, Methods in Molecular Biology, vol. 1106,
DOI 10.1007/978-1-62703-736-5_19, © Springer Science+Business Media, LLC 2014

bacterial colonization [4, 7]. In accordance with other reports, our studies demonstrated the requirement for immunosuppression to establish *S. epidermidis* infection in an in vivo rat model [5, 6].

2 Materials

2.1 Catheter Construction

1. Tygon® S-54-HL Microbore Tubing, 0.3 mm ID × 0.8 mm OD (catalog number 720993) (Harvard Apparatus, Holliston, MA).
2. Silicone sealant.
3. Twenty-six gauge × 5/8″ (Sub-Q) needle.
4. Three milliliter or 5 mL syringe.
5. Razor blade.
6. Sterilization pouches (catalog number 024008) (Propper Manufacturing Co., Inc., Long Island City, NY).

2.2 Catheter Placement Surgery

1. Sterile drapes.
2. Sterile swabs or non-adhering pads.
3. Puralube® vet ointment or comparable ophthalmic ointment.
4. Povidone–Iodine cleansing solution, 7.5 %.
5. Heating pad(s).
6. Electric trimmer.
7. Washcloth or towel.
8. Scale.
9. Twenty-six gauge × 5/8″ (Sub-Q) needles.
10. One milliliter syringes.
11. Ketamine HCl concentrate, 100 mg/mL.
12. Xylazine, 100 mg/mL in sterile water.
13. Cefazolin, 300 mg/mL in sterile saline.
14. Buprenex® buprenorphine HCl, 0.3 mg/mL (dilute 1:10 in sterile saline or water) (Reckitt Benckiser Pharmaceuticals Inc., Richmond, VA).
15. Catheters.
16. 3-0 Prolene, monofilament suture with needle.
17. 4-0 Silk, braided suture ties.
18. Scalpel with #15 blades.
19. Iris scissors, straight.
20. Jeweler's forceps, straight tip, sharp.
21. Fell needle holder.
22. Delicate Mosquito forceps.

23. Castroviejo Micro Scissors, curved blade, round handle.
24. Oschner Forceps, 90° jaw.

2.3 Immuno-suppression

1. Cyclophosphamide monohydrate (CP) (Catalog number 150749) (MP Biomedicals LLC, Solon, OH): Prepare a 40 mg/mL solution in sterile saline. Store at −20 °C (*see* **Note 1**).
2. Washcloth or towel.
3. Twenty-six gauge × 5/8″ (Sub-Q) needles.
4. One milliliter and 3 mL syringes.
5. Scale.

2.4 Bacterial Inoculation and Enumeration

1. Tryptic Soy Broth (TSB).
2. One liter flask(s).
3. Centrifuge and appropriate tubes (*see* **Note 2**).
4. Sterile saline or phosphate buffered saline (PBS).
5. Dilution tubes.
6. Agar plates.
7. DecapiCone® Disposable Rodent Restrainers (catalog number DC-200) (Braintree Scientific Inc., Braintree, MA).
8. Medium binder clips (1¼ in.).
9. Heating pad(s).
10. Alcohol prep pads, 70 % isopropyl alcohol.
11. Twenty-six gauge × 5/8″ (Sub-Q) needles.
12. One milliliter syringes.
13. 1.5 mL tubes containing 1 mL sterile PBS with 0.5 % Tween-20 (*see* **Note 3**).
14. Blood collection tubes containing anticoagulant (*see* **Note 4**).
15. Fifty milliliter conical tubes.
16. Fourteen milliliter round-bottom tubes.
17. Tissue homogenizer (*see* **Note 5**).

3 Methods

3.1 Catheter Preparation

1. Prepare catheters by cutting 7.5 cm lengths of tubing. Inject silicone sealant into one end of the tubing segments using a 26 gauge needle attached to a 3 mL or 5 mL syringe; occlude 2–2.5 cm of the tubing.
2. After the sealant has cured, use a clean blade to cut a bevel on the sealed end. Place catheters individually into sterilization pouches and sterilize with ethylene dioxide gas.

3.2 Insertion of Jugular Catheters

1. Weigh rats to determine anesthetic and analgesic dosage amounts (*see* **Note 6**).
2. Anesthetize rats with an intraperitoneal (IP) injection of ketamine and xylazine (*see* **Note 7**).
3. As a prophylactic measure, inject 100 μL of cefazolin intramuscularly (IM) in the quadriceps.
4. Inject diluted buprenorphine (0.01–0.05 mg/kg) subcutaneously (SQ) for analgesia. Repeat every 8–12 h if signs of pain persist or as directed by your institution's IACUC.
5. Use the electric trimmer to shave the right ventral side of the throat.
6. Apply ophthalmic ointment (*see* **Note 8**).
7. Place the rat in a supine position, securing the limbs with soft restraints. Apply povidone–iodine solution to the shaved area using sterile swabs or non-adhering pads.
8. Make an approximately 2.5 cm incision lateral to the trachea, over the jugular vein, using a #15 blade scalpel. Bluntly dissect away the soft tissue from the jugular vein using the mosquito forceps, iris scissors, and jeweler's forceps. After isolating the vein, place a 4-0 silk suture posterior to it using the Oschner forceps (Fig. 1).
9. Make a small transverse venotomy (no greater than ½ the diameter of the vein) with the curved Castroviejo micro scissors at the level of the silk suture located behind the jugular vein.
10. Insert the beveled end of a catheter into the venotomy and advance it into the superior vena cava.

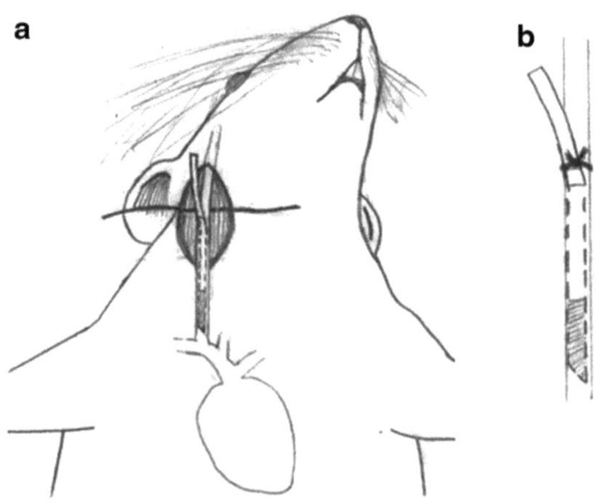

Fig. 1 Diagram of catheter localization within the external jugular vein. (**a**) The beveled end of the catheter is advanced into the superior vena cava, then (**b**) secured with silk suture

11. Secure the catheter with the 4-0 silk tie that was previously passed around the jugular vein. Position the extravascular portion of the catheter in the soft tissue of the neck. Close the incision with running stitch using the 3-0 monofilament suture.
12. Reapply ophthalmic ointment as necessary and remove excess blood using alcohol prep pads. Keep animals warm (using heating pads) until they awake.

3.3 Immunosuppression by Cyclophosphamide (CP)

1. Four days after catheter insertion, inject rats with 100 mg/kg CP, IP (*see* **Note 9**).
2. Three days later, inject rats with 50 mg/kg CP, IP (*see* **Note 10**).

3.4 Bacterial Preparation and Inoculation

1. On the day of the second CP injection, inoculate 100 mL of TSB in a 1 L flask. Incubate at 37 °C with agitation for 18 h.
2. The following day, spin down 30 mL of the culture at 6,000 × *g* for 3 min (*see* **Note 11**).
3. Decant the supernatant and wash the cell pellet with 10–15 mL sterile saline or PBS. Spin as before.
4. Decant the supernatant and resuspend the cell pellet in 3 mL sterile saline or PBS (*see* **Note 12**).
5. Ensure that the rats are securely restrained. Use heating pads to warm rats and bring tail vasculature to the surface (*see* **Note 13**).
6. Wipe tails with an alcohol prep pad before injecting 100 μL of the prepared inoculum into either lateral tail vein using a 26 gauge needle.
7. Three days after inoculation, sacrifice rats by carbon dioxide inhalation and aseptically harvest the catheter (and if desired, the blood, liver, and spleen) for bacterial enumeration (*see* **Note 14**).

4 Notes

1. The solubility of CP in aqueous solution is 40 g/L. Unused CP may be stored at −20 °C, but comes out of solution; to resolubilize, warm the CP to 25–35 °C.
2. Any sterile tubes able to hold at least 30 mL will work. We use 50 mL Nalgene Oak Ridge Centrifuge Tubes (Nalge Nunc International, Rochester, NY) with an SS-34 fixed angle rotor in a Sorvall RC 5C PLUS centrifuge (Thermo Fisher Scientific, Waltham, MA).
3. To process catheters, place each into a 1.5 mL tube containing 1 mL sterile PBS with 0.5 % Tween-20. This solution aides bacterial removal but does not affect viability. Vortex tubes for

1 min, sonicate for 5 min, then vortex again for 1 min. Dilution plate the sonicate using standard procedures.

4. Collect blood in tubes containing anticoagulant; based on the findings of Evans et al. we use heparin, but EDTA is also suitable when collecting blood for the purpose of dilution plating [8].
5. To enumerate the bacterial load in organs it is necessary to homogenize the tissues; we use a PowerGen 700D (Fisher Scientific, Pittsburgh, PA) electric homogenizer for this purpose.
6. Anesthetic dosage for a rat is 40–87 mg/kg ketamine and 5–13 mg/kg xylazine. For an approximately 300 g rat, we use 300–350 μL ketamine (100 mg/mL) and 30 μL xylazine (100 mg/mL). Keep animals warm on a heating pad (set to low) if anesthetizing more than one animal at a time.
7. When performing IP injections, we use a variation of the "rat wrap" method [9]. Place the rat at one corner of a washcloth or small towel. While securing the rat, fold the remaining three corners of the cloth over the rat as if folding a burrito.
8. We perform **steps 7–11** of the catheter placement surgery in a laminar flow hood.
9. Animals should be allowed to recover for at least 4 days after surgery, and it is fine to wait additional days before beginning CP treatment. However, once CP treatment is started the dosing and inoculation schedule should be followed as outlined.
10. We use the same 40 mg/mL CP stock solution and simply halve the dosage volume.
11. Bacterial culture(s) should be approximately 10^9 CFU/mL. The centrifugation steps concentrate the culture(s), resulting in an inoculum of approximately 10^9 CFU in 100 μL.
12. Dilution plate to confirm the inoculum. Keep bacterial preparation(s) on ice until use.
13. For IV injections, we use DecapiCone Disposable Rodent Restrainers with a slit along the seam (at the large end of the cone) to allow access to the tail. Keep rats securely in the cone using medium binder clips (usually 3/cone) (Fig. 2). If cones are too large, trim excess length from the base to ensure a more "snug" fit; immobilizing the rats is the key to successful IV injections. Additionally, to confirm needle placement, gently pull back on the syringe before injecting bacteria. There should be a flash of blood in the needle hub when the needle is in the vein. For additional information on the tail vasculature and recommendations for injections see Staszyk et al. [10].
14. During harvesting, place organ(s) of interest in pre-weighed tubes; collect livers in 50 mL tubes and spleens in 14 mL tubes.

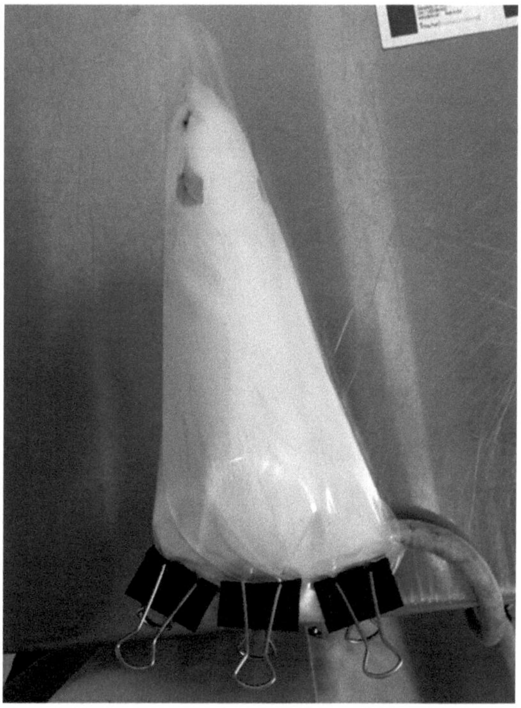

Fig. 2 Rat restrained for tail vein inoculation. To access the lateral tail veins rats are immobilized using DecapiCone Disposable Rodent Restrainers secured with binder clips

After harvesting, re-weigh tubes and then calculate the tissue weight. For large organs, such as the liver, add 1.5 mL sterile saline or PBS per gram of tissue before homogenizing. Colony-forming units per gram (CFU/g) of tissue is obtained by dividing the colony counts by 0.67. For smaller tissues, such as the spleen, add 1 mL sterile saline or PBS to tubes before homogenizing. Divide colony counts by organ weights to determine CFU/g of tissue.

References

1. NNIS (2004) National Nosocomial Infections Surveillance (NNIS) System Report, data summary from January 1992 through June 2004, issued October 2004. Am J Infect Control 32: 470–485
2. Fey PD (2010) Modality of bacterial growth presents unique targets: how do we treat biofilm-mediated infections? Curr Opin Microbiol 13:610–615
3. Rupp ME, Ulphani JS, Fey PD et al (1999) Characterization of Staphylococcus epidermidis polysaccharide intercellular adhesin/hemagglutinin in the pathogenesis of intravascular catheter-associated infection in a rat model. Infect Immun 67:2656–2659
4. Mehall JR, Saltzman DA, Jackson RJ et al (2002) Fibrin sheath enhances central venous catheter infection. Crit Care Med 30:908–912
5. Ebert T, Smith S, Pancari G et al (2011) Development of a rat central venous catheter model for evaluation of vaccines to prevent *Staphylococcus epidermidis* and *Staphylococcus aureus* early biofilms. Hum Vaccin 7:630–638
6. Chauhan A, Lebeaux D, Decante B et al (2012) A rat model of central venous catheter to study establishment of long-term bacterial

biofilm and related acute and chronic infections. PLoS One 7:e37281

7. Keller JE, Hindman JW, Mehall JR et al (2006) Enoxaparin inhibits fibrin sheath formation and decreases central venous catheter colonization following bacteremic challenge. Crit Care Med 34:1450–1455
8. Evans GL, Cekoric T Jr, Searcy RL (1968) Comparative effects of anticoagulants on bacterial growth in experimental blood cultures. Am J Med Technol 34:103–112
9. Sharp PE, La Regina MC (1998) Experimental methodology. In: Suckow MA (ed) The laboratory rat. CRC Press, Boca Raton, pp 129–159
10. Staszyk C, Bohnet W, Gasse H et al (2003) Blood vessels of the rat tail: a histological reexamination with respect to blood vessel puncture methods. Lab Anim 37:121–125

Index

Paul D. Fey (ed.), *Staphylococcus Epidermidis: Methods and Protocols*, Methods in Molecular Biology, vol. 1106, DOI 10.1007/978-1-62703-736-5, © Springer Science+Business Media, LLC 2014

MIX
Papier aus verantwortungsvollen Quellen
Paper from responsible sources
FSC® C105338

If you have any concerns about our products,
you can contact us on
ProductSafety@springernature.com

In case Publisher is established outside the EU,
the EU authorized representative is:
Springer Nature Customer Service Center GmbH
Europaplatz 3, 69115 Heidelberg, Germany

Printed by Libri Plureos GmbH
in Hamburg, Germany